U0162178

都匀毛尖茶

DUYUN MAOJIAN TEA

沈苏文 著

吉林出版集团股份有限公司

图书在版编目（CIP）数据

画说都匀毛尖茶 / 沈苏文著 . — 长春 : 吉林出版集团股份有限公司 , 2020.10

ISBN 978-7-5581-9249-4

Ⅰ . ①画… Ⅱ . ①沈… Ⅲ . ①绿茶—茶文化—都匀—图解 Ⅳ . ① TS971.21-64

中国版本图书馆CIP数据核字(2020)第187711号

画说都匀毛尖茶

著　　者	沈苏文
责任编辑	齐　琳　史俊南
责任校对	董　凯
封面设计	邢海燕
开　　本	787mm × 1092mm　1/16
字　　数	279千字
印　　张	14.5
版　　次	2020年10月第1版
印　　次	2020年10月第1次印刷
出　　版	吉林出版集团股份有限公司
电　　话	总编办：010—63109269
	发行部：010—85173824
印　　刷	河北盛世彩捷印刷有限公司

ISBN 978-7-5581-9249-4　　定价：98.00 元

中国名茶
China's Famous Tea

PREFACE

序

庚子年仲春，到都匀市政协调研，获悉有人用诗画形式阐释都匀毛尖茶，作为一名茶人，我心甚慰。未曾想到，不到半年，《画说都匀毛尖茶》集成付印，嘱我作序。如此善举，该当点赞，便欣然为之。

中国文人常常把“琴棋书画诗曲茶”作为生活中七大雅好，书画诗茶融合唱和，能够很好诠释文化、艺术和生活的真谛，彰显“和”“静”“怡”“真”的中国茶道精神。

茶道，非常道！唐永泰元年（765年），陆羽《茶经》问世，开启了茶与文化交融之门，1200多年来，茶艺、茶礼、茶境至成至臻。唯其如此，得益于文化的滋润涵养，茶道更显博大精深。

卢仝《走笔谢孟谏议寄新茶》云：“一碗喉吻润，二碗破孤闷。三碗搜枯肠，唯有文字五千卷。四碗发轻汗，平生不平事，尽向毛孔散。五碗肌骨清，六碗通仙灵。七碗吃不得也，唯觉两腋习习清风生。”咏赞字句，脍炙人口，千年不衰。北宋范仲淹描述文人雅士品茗论水、斗茶和煮茶技艺的《和项目岷从事斗茶歌》，意境妙趣横生，茶文化意蕴浓厚；通晓音律、工于书画的宋徽宗赵佶钦笔之作《大观茶论》，定格了茶文化的一个精彩片断。

都匀毛尖不仅金玉其外，而且内秀于中。查阅志书史籍，徜徉于历史长廊，我们发现，都匀毛尖茶有着厚重的文化底蕴。

1915年，都匀毛尖茶远渡重洋，在大西洋彼岸的旧金山荣获了巴拿马博览会金奖。1982年8月，都匀毛尖茶跻身全国十大名茶之列。2010年，都匀毛尖茶荣登上海世博会十大名茶宝座。

沈苏文先生满怀对都匀毛尖茶的深情厚意，以自己独到的视野，从2018年开始精心收集、整理、挖掘都匀毛尖茶相关史料，广泛结交诗书画名家，借助大家的妙手丹青，量身打造都匀毛尖茶文化与地方历史文化、民族文化相互交融的书画作品，“以茶入画，以诗释茶”，用雅俗共赏的艺术形式，大力彰显都匀毛尖茶超凡脱俗的高贵品质和都匀毛尖茶深邃的文化内涵，独辟蹊径，自成风格，匠心可嘉。

“读图时代”是一个快节奏生活的时代，也是一个信息极速传播的时代，我们希望，您走进了《画说都匀毛尖茶》的世界，都匀毛尖茶就走进了您的世界。

2020年仲秋　于杉木湖畔

用名家画笔

为都匀毛尖茶立传

With the paintbrushes of famous painters,

I made a biography of Duyun Maojian tea.

C O N T E N T S

目 录

CHAPTER 1
Chinese Treasures
第一篇 华夏瑰宝

CHAPTER 2
The Capital of National Tea
第二篇 国茶之都

CHAPTER 3

Famous for Thousands of Years

第三篇　名垂千古

一　毛尖与名人

二　毛尖茶史

三　毛尖传说

CHAPTER 4
Brand Creation
第四篇　品牌缔造

一　毛尖醉美

二　毛尖茶诗

三　毛尖茶人

四 毛尖远航

CHAPTER 5

Tea Painting Forum

第五篇　茶画春秋

一　名家说茶

二　问道茗香

三　茶来茶去

四　茶语清心

Tea is the home of the soul.

Painting is the companion of the soul.

Let tea enter the painting and explain tea with painting.

The relationship between tea and painting is worthy of being highlighted in the history of Chinese tea culture.

The beauty of tea and painting is the spiritual understanding captured by the thousands of years of Chinese tea people's life practice.

Su Dongpo said that although there are different forms of tea and painting, their connotation and beauty are interlinked.

It is: those who are really addicted to alcohol are valiant, and those who are really addicted to tea are refreshing.

茶，是灵魂的家。

画，是灵魂的伴。

以茶入画，以画释茶。

茶与画的结缘，在中国茶文化史上值得用浓墨重彩大书特书。

茶与画的妙得，是中国茶人数千年来在实践中捕获的心灵感悟。

苏东坡曰：“上茶妙墨俱香，是其德同也；皆坚，是其操同也，譬如贤人君子黔晰美恶之不同，其德操一也。”

正是，真嗜酒者气雄，真嗜茶者神清。

Chapter I
Chinese Treasures

A beautiful tree carries five thousand years and contains five thousand years of cultural genes. Since ancient times, tea and culture have made an indissoluble bond. It can be said that there is no cultural attachment, tea is just a leaf. Besides its noble quality, the real tea is not only accidental, but also inevitable to be associated with culture.

第一篇 华夏瑰宝

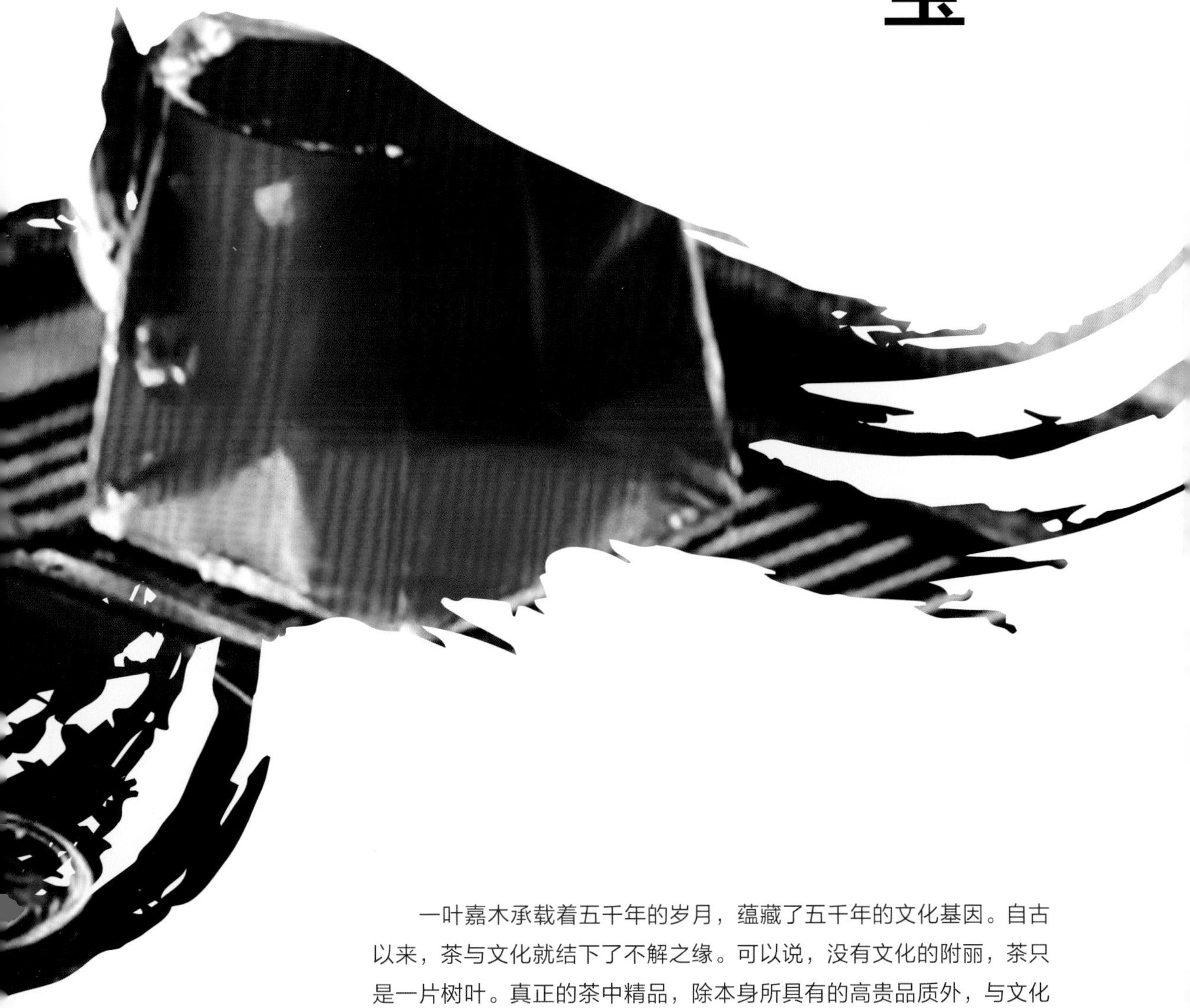

一叶嘉木承载着五千年的岁月，蕴藏了五千年的文化基因。自古以来，茶与文化就结下了不解之缘。可以说，没有文化的附丽，茶只是一片树叶。真正的茶中精品，除本身所具有的高贵品质外，与文化的结缘，既是偶然，也是必然。

茶（作者：仲冰）

经典 1

“茶者，南方嘉木也。”（唐·陆羽《茶经》）

一片树叶，惠及众生。当今世界，有 60 多个国家和地区种植茶，遍布世界五大洲。有 160 多个国家和地区有饮茶习俗，有 30 多亿人钟情茶饮，茶已成为世界上仅次于水的健康饮料。

A leaf benefits all living beings. Today, there are more than 60 countries and regions planting tea in the world, covering five continents. There are more than 160 countries and regions in the world that have tea drinking customs. More than 3 billion people are in love with tea. Tea has become the world's second only to water healthy drinks.

经典2

饮茶起源说

中国饮茶起源神农说（作者：王殿维 临）

“茶之为饮，发乎神农氏。”（唐·陆羽《茶经》）中国的饮茶传统源于神农氏的说法因民间传说而衍生出了不同的观点。有人认为茶是神农在野外以釜锅煮水时，刚好有几片叶子飘进锅中，煮好的水，其色微黄，入口甘甜止渴、提神醒脑，以神农尝百草的经验，判断它是一种药而被发现的。这是有关中国饮茶的起源最普遍的一种说法。

“Tea as a drink was discovered by Shennong.” (Tea Classics by Lu Yu in the Tang Dynasty) The idea that Chinese tea drinking originated from Shennong derived different opinions from folklore. Some people thought that when Shennong cooked water in a pot in the wild, there were just a few leaves floating in to the pot. The boiled water had a slight yellow color. It was sweet, thirst quenching and refreshing to drink in the mouth. Based on Shennong's experence of tasting hundreds of herbs in the past, it would be judged that it was a kind of medicine. This is the most common statement about the origin of Chinese tea drinking.

茶香琴韵（作者：傅宝）

经典3

茶的前世今生

中国是世界上最早发现茶、种植茶、饮用茶的国家，是世界茶文化的发祥地。中国茶的发现和利用已有4700多年的历史，长盛不衰，传遍全球。茶是中华民族的举国之饮，发于神农氏，闻于鲁周公，兴于唐朝，盛于宋代，普及于明清之际，发展于当今盛世。

China is the first country in the world to discover, grow and drink tea, and the birthplace of world tea culture. The discovery and utilization of Chinese tea has a history of more than 4700 years, and it has been flourshing all over the world. Tea is the national drink of the Chinese nation. It originated in Shennong, became famous since Lu Zhougong, rose in Tang Dynasty, flourished in Song Dynasty, popularized in Ming and Qing Dynasties, and develop in the present flourishing age.

从来佳茗似佳人（作者：舒肖）

茶是中华历史文化中的瑰宝，也是中国的文化名片。茶之于中国，就如同红酒之于法国，啤酒之于德国。她是中国的骄傲，是民族的自尊、自信和自豪。世界著名科技史学家李约瑟博士，将中国茶叶作为中国继“四大发明”（火药、造纸术、指南针和印刷术）之后，对人类的第五个重大贡献。

Tea is not only a treasure in Chinese history and culture, but also a cultural card of China. Tea to China is like wine to France and beer to Germany. She is the pride fo China, the national self-esteem, self-confidence and glory. Dr. Joseph Needham, a famous historian of science and technology in the world, took Chinese tea as the fifth great contribution to mankind after the four great inventions of China (gunpowder, papermaking, compass and printing).

心上毛尖（作者：舒肖 临）

经典5 中国茶文化

种茶、饮茶是前提，加上文人的参与和文化的内涵，才能真正形成茶文化。唐代陆羽和皎然等一大批文人非常重视茶的精神享受和道德规范，推崇饮茶用具、饮茶用水和煮茶艺术，并与儒、道、佛哲学思想相互交融，还在饮茶过程中创作了很多“茶诗”“茶画”，从而奠定了中国茶文化的基础。

Planting tea and drinking tea are the premise. Only with the participation of scholars and the connotation of cultrue, can tea cultre be truly formed. Lu Yu and Jiaoran, a large number of cultural people in Tang Dynasty, attached great importance to the spiritual enjoyment and moral standards of tea, paid attention to tea drinking utensils, tea drinking water and tea making art, mingled with Confucianism, Taoism and Buddhism, and created many tea poems and tea paintings in the process of tea drinking, which laid the foundation of Chinese tea culture.

经典6 中国茶德

从文化的角度看，茶并不只是一片树叶，也不只是一种饮料。它富有博大精深的内涵，源远流长的历史，并兼具物质属性和精神属性两大特性，具有政治、经济、社会、文化、生态、养生等多种功能。中国国际茶文化研究会提炼出当代茶文化核心理念，即中国茶德："清、敬、和、美"。清为本，敬为上，和为核，美为境。

From the cultural point of view, tea is not just a leaf or a drink. It is rich in profound connotation and has a long history. It has both material and spiritual characteristics, and has many functions such as politics, economy, society, culture, ecology, and health preservation. China International Tea Culture Research Association abstracts the core concept of contemporary tea culture: "purity, respect, harmony and beauty". The purity is the foundation, the respect is the best, the harmony is the core, and the beauty is the realm.

高士雅集图（作者：贺成才）

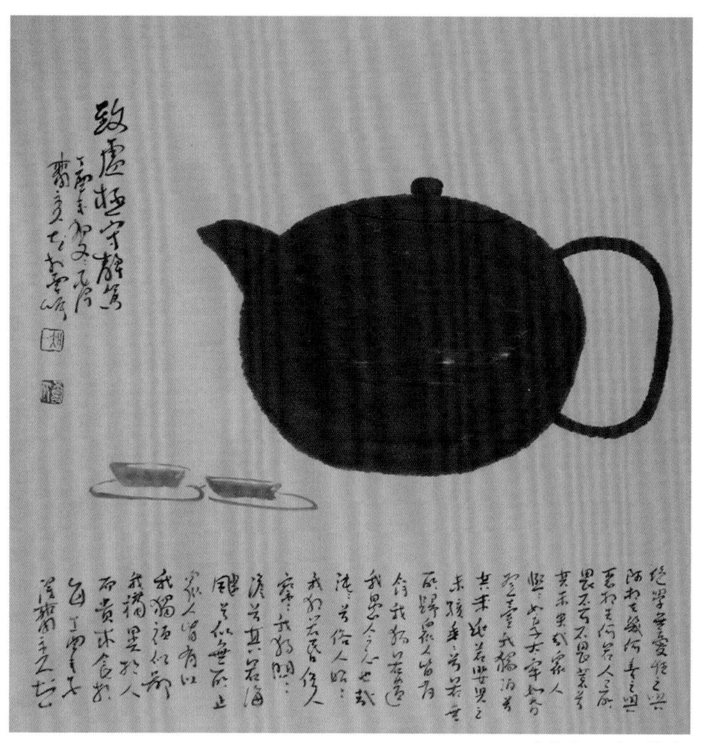

壶（作者：李云峰）

经典 7

中国茶道

茶道，就是品赏茶的美感之道，亦被视为一种烹茶、饮茶的生活艺术，一种以茶为媒的生活礼仪，一种以茶修身的生活方式。它通过沏茶、赏茶、闻茶、饮茶等，增进友谊，美心修德，学习礼法，领略传统美德，是一种很有益的和美仪式。茶道最早源于中国，中国人至少在唐代或唐代以前，就在世界上首先将茶饮作为一种修身养性之道。

Sado is the aesthetic way of appreciating tea. It is also regarded as a life art of tea making and drinking, a life etiquette with tea as the medium, and a life style of self-cultivation with tea. It is a very beneficial, harmonious and beautiful ceremony to enhance friendship, cultivate virtue, learn etiquette and appreciate traditional virtue by making, enjoying, smelling and drinking tea. Tea ceremony originated in China. Before Tang Dynasty or Tang Dynasty at least, Chinese people first took tea drinking as a way of self-cultivation in the world.

经典 8

中国茶礼

通常，传统意义上的茶礼，指古时候男方向女方下聘，以茶为礼，称为“茶礼”，又叫“吃茶”。古时因茶树移植则不生，种树必下籽，故在古代婚俗中，茶便成为坚贞不移和婚后多子的象征，婚娶聘物必定有茶。

Usually, tea gift was an engagement gift given by ancient Chinese men when they propose to women in the traditional sense. It was also called "eating tea". In ancient times, tea trees could not be transplanted, and planting tea trees was bound to produce seeds. Therefore, in ancient marriage customs, tea became a symbol of constancy and having many children after marriage. There must be tea in the dowry.

中国茶礼（作者：尚秋香）

以茶会友（作者：尚秋香）

经典 9

中国茶艺

茶艺是饮茶活动中特有的文化现象，是一门唯美是求的生活艺术，包括选茗、择水、烹茶技术、茶具艺术、环境布置，等等。茶艺萌芽于唐，发扬于宋，改革于明，极盛于清，可谓有相当的历史渊源，自成一统。

Tea art is a unique cultural phenomenon in tea drinking activities. It is a kind of life art with aesthetic pursuit, including tea selection, water selection, tea cooking technology, tea set art, environmental layout and so on. Tea art sprouted in the Tang Dynasty, developed in the Song Dynasty, reformed in Ming Dynasty and flourished in the Qing Dynasty. It can be said that it has quite a historical origin and has its own unity.

经典10 中国贡茶

古韵图（作者：马纪奎）

最初的贡品是土贡方物，所贡之物皆为本地土特产。虽然没有文献记载，但贡茶至少在西汉时即已出现。贡茶因制作精致至极而价格高昂，与金等价，甚至金可得而贡茶不可得。民国初年，中华民国政府宣布停止各省向中央政府贡茶，从此废除了千百年来的贡茶制度。

The original tribute was a local tribute, that was, the tribute was a local specialty. Although there was no literature, tribute tea appeared at least in the Western Han Dynasty. Tribute tea was extremely expensive because of its exquisite production. It was as expensive as gold. Even gold was available, but tribute tea was not available. In 1912, the government of the Republic of China announced that the provinces would stop offering tea to the central government, and the system of tribute tea for thousands of years was abolished.

读书品茗（作者：墨梦 临）

经典 11

中国四大茶区

中国作为茶叶大国，茶区东起台湾东部海岸，西至西藏贡茶场，南至海南岛榆林海港，北到山东荣成，包含西南茶区、华南茶区、江南茶区、江北茶区四大茶区。西南茶区是中国最古老的茶叶产地，包括云南、贵州、四川、西藏东南部地区，中国的外销碎茶和边茶大多产自这里。

As a big tea country, China's tea area starts from the east coast of Taiwan in the East, west to the tribute tea farm of Tibet, south to the Yulin seaport of Hainan, north to the Rongcheng of Shandong, including four tea areas: Southwest tea area, South China tea area, Jiangnan tea area and Jiangbei tea area. Southwest tea area is the oldest tea producing area in China, including Yunnan, Guizhou, Sichuan and Southeast Tibet. Most of China's export broken tea and side tea are produced here.

经典 12

中国六大茶类

我国茶叶分类目前尚未有统一的方法。安徽农业大学陈椽教授提出，按制法和品质为基础，以茶多酚氧化程度为序，把初制茶叶分为绿茶、黄茶、黑茶、青茶、白茶、红茶等六大类。以都匀茶叶专家的说法，结合茶叶的商品形态和冲泡形式，广义茶叶种类还应包括花茶和非茶之茶，共八大类。平时经常说的大麦茶、荞麦茶、药茶等，虽以茶相称，但实为另外的东西，只是冲泡和品饮方式近似而已，所以称为“非茶之茶”。

There is no unified method for tea classification in China. Professor Chen Chuan of Anhui Agricultural University proposed that primary tea should be divided into six categories, namely green tea, yellow tea, dark tea, Cyan tea, white tea and black tea, based on the processing method and quality, and taking the oxidation degree of tea polyphenols as the order.

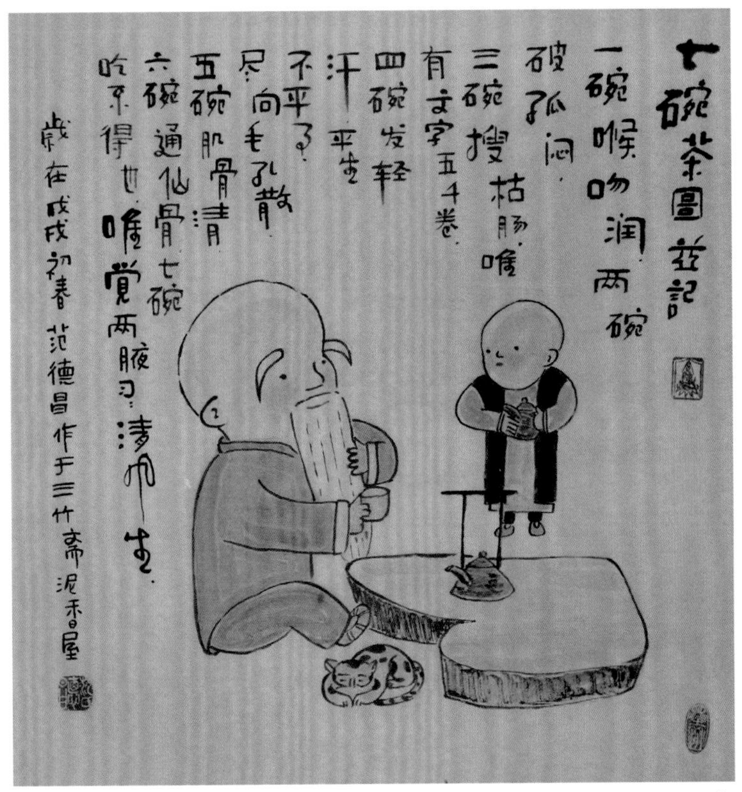

七碗茶图（作者：范德昌）

经典 13

中国十大名茶

中国茶文化已经传承了几千年，叫得上名字的茶就有上千种。不同时期、不同评选机构出炉的中国十大名茶也有些许出入，但普遍承认的中国十大名茶是：西湖龙井、洞庭碧螺春、都匀毛尖、黄山毛峰、六安瓜片、信阳毛尖、安溪铁观音、武夷岩茶、祁门红茶、君山银针。

Chinese tea culture has been passed down for thousands of years. There are thousands of tea that can be named. There are also some differences among the top ten famous Chinese teas selected by different organizations in different periods, but the commonly recognized top ten are: West Lake Longjing, Dongting Biluochun, Duyun Maojian, Huangshan Maofeng, Lu'an Guapian, Xinyang Maojian, Anxi Tieguanyin, Wuyi rock tea, Qimen black tea, Junshan silver needle.

清茶一杯也醉人（作者：尚秋香）

经典 14

茶的古丝绸之路

传道解惑（作者：白林）

西汉时，著名外交家张骞奉汉武帝之遣，先后两次出使西域，著名的丝绸之路随之开通。这条横贯欧亚的商贸通道今被称为“古丝绸之路”。接着，中国的茶叶连同丝绸、瓷器等货物源源不断地从西安出发，经甘肃、过新疆，直达中亚、西亚、南亚，直至欧洲等许多国家，后遍及欧亚大陆。

In the Western Han Dynasty, Zhang Qian, a famous diplomat, was sent to the western regions twice by Emperor Wu of the Han Dynasty. The famous Silk Road in Chinese history was opened. This trade passage across Europe and Asia is now known as the "ancient Silk Road". Then, China's tea, silk, porcelain and other goods continue to flow from Xi'an, through Gansu and Xinjiang, to Central Asia, West Asia, South Asia, Europe and many other countries. Finally, it covers Eurasia.

经典 15

茶的海上丝绸之路

汉代以后，特别是隋唐时期，随着以宁波、泉州、广州为起点的“海上丝绸之路”的贯通，茶叶又随同丝绸、瓷器等货物从东临的朝鲜半岛、日本开始远涉重洋，直至美洲、非洲、大西洋和世界上的其他国家。

After the Han Dynasty, especially the Sui and Tang Dynasties, with the opening of the “Maritime Silk Road” starting from Ningbo, Quanzhou and Guangzhou, tea, along with silk, porcelain and other goods, began to cross the ocean from the east to the Korean Peninsula, Japan, until the Americas, Africa, the Atlantic Ocean and other countries of the world.

品茶悟道（作者：白林）

茶圣陆羽图（作者：王殿维 临）

经典 16

茶圣陆羽

唐代陆羽，湖北天门人，一生只做两件事：品茶和记录。1200 多年前，他将自己毕生饮茶的经验心得记录成一本书，这就是世界上第一部茶叶专著——《茶经》。此书一出，轰动全唐。从此，中国人家家饮茶。

Lu Yu, a famous tea expert in Tang Dynasty and a native of Tianmen, Hubei province, did only two things in his life: tea tasting and recording. More than 1200 years ago, he wrote a book about his whole life's tea drinking experience, which is the world's first tea monograph, the Book of Tea. As soon as this book was published, it made a sensation in the whole Tang Dynasty. From then on, Chinese people drink tea.

茶别人白居易图（**作者：王殿维 临**）

经典 17

“别茶人”白居易

唐代极负盛名的大诗人白居易，也是一位很有品位的茶客，自称“别茶人”（鉴别茶叶的人）。他一生爱诗、嗜酒、痴茶、好琴，即使暮年之际，茶、酒、老琴依然是与他长相左右的莫逆知己，并留下了许多与茶相关的优美诗句。

Bai Juyi, a famous poet in Tang Dynasty, is also a tea man with good taste. He called himself “Biecha man” (the person who identifies tea). He loved poetry, wine, tea and Qin all his life. Even in his old age, tea, wine and old Qin were still his constant confidants, and he left many beautiful poems related to tea.

经典 18 『茶神』陆游

南宋著名的爱国诗人陆游，也是南宋著名的茶人。他生于茶乡，做过茶官，晚年又归隐茶乡，对茶倾注了无限深情。陆游与茶的关系，正像陶渊明与菊、李白与酒一样的深切。他爱茶如诗，研读《诗经》，崇拜陆羽，道出了“桑苎家风君勿笑，他年犹得作茶神”的精神向往。

Lu You was not only a famous patriotic poet in the Southern Song Dynasty, but also a famous tea man in the Southern Song Dynasty. He was born in the tea country and served as a tea official. In his later years, he returned to the tea country and devoted himself to tea. The relationship between Lu You and tea is just as profound as that between Tao Yuanming and Chrysanthemum, Li Bai and wine. He loved tea as poetry, studied the Book of Tea, and worshipped Lu Yu.

茶神陆游图（作者：王殿维 临）

经典 19

『茶仙』苏东坡

茶仙苏东坡图（作者：王殿维 临）

我国宋代杰出的文学家苏东坡，在文、诗、词三方面都达到了极高的造诣，堪称宋代文学最高成就的代表。他十分嗜茶。茶，助诗思，战睡魔，是他生活中不可或缺之物。他对品茶、烹茶、种茶样样在行，对茶史、茶功颇有研究，还创作出众多咏茶诗词。“酒困路长惟欲睡，日高人渴漫思茶，敲门试问野人家。”这几句诗形象地记述了他讨茶解渴的情景。

Su Dongpo was an outstanding litterateur in the Song Dynasty in China. He had achieved great achievements in literature, poetry and CI, which can be regarded as the representative of the highest literary achievements of Song Dynasty. He was very fond of tea. Tea, was indispensable in his life, helping poetry and thinking, and fighting against sleeping demons. He was good at tea tasting, tea cooking and tea planting. He had a lot of research on the history and efficacy of tea. He also created many poems on tea.

经典20

『茶怪』郑板桥

茶怪郑板桥图（作者：王殿维 临）

清代官吏、书画家、文学家，“扬州八怪”之一的郑板桥，一生主要客居扬州，以卖画为生。其诗、书、画均旷世独立，世称“三绝”，擅画兰、竹、石、松、菊等植物，其中画竹逾五十余载，成就最为突出。郑板桥是个茶人，他写了很多茶诗和茶词，是历代写茶联最多的一个茶人。著名的《竹枝词》云：“溢江江口是奴家，郎若闲时来吃茶。黄土筑墙茅盖屋，门前一树紫荆花。”

Zheng Banqiao, one of the "Magical Eight Painters of Yangzhou", was an official, calligrapher, painter and litterateur in Qing Dynasty. In his life, he lived mainly in Yangzhou and made a living selling paintings. His poems, calligraphy and paintings were all independent from the world, which was known as "three wonders". He was good at painting orchids, bamboos, stones, pines, chrysanthemums and other plants, of which bamboos had been painted for more than 50 years, with the most outstanding achievements. Zheng Banqiao was a tea man. He wrote many tea poems and CI. He was the tea man who wrote the most tea couplets in all dynasties.

经典21

中国茶征服了全世界

山茶花（作者：周娟）

在世界三大饮料（茶、咖啡、可可）中，只有茶叶成功地征服了全世界。十七世纪，中国茶叶的出口量就已超过瓷器和丝绸，约占出口货物的90%。据说，茶叶是有史以来最先在伦敦做广告的商品。"中国风"不仅席卷了英伦，还风靡了世界。从十七世纪到十八世纪，全世界都流行"吃茶去"。如今，茶叶在世界各地无处不在，茶叶的消费超过了咖啡、巧克力、可可、碳酸饮料和酒精饮料的总和。

Among the three major drinks in the world(tea, coffee and cocoa), only tea has successfully conquered the whole world. In the 17th century, China's export of tea exceeded that of porcelain and silk, accounting for about 90% of the export goods. It is said that tea is the first product ever advertised in London. "Chinese style" not only swept UK, but also swept the world. From the 17th century to the 18th century, tea was popular all over the world. Nowadays, tea is everywhere in the world, and the consumption of tea in the world exceeds the sum of coffee, chocolate, cocoa, carbonated drinks and alcoholic drinks.

Chapter II
The Capital of National Tea

Duyun City, with the title of "Landscape Bridge City, the Capital of National Tea", has beautiful mountains and rivers and rich ethnic customs. It is the capital of Qiannan Buyi and Miao Autonomous Prefecture in Guizhou Province, the excellent tourist city in China and the global green city. The total area of the city is 2274 square kilometers, with a total population of 500,000. There are 33 ethnic minorities, such as Buyi, Miao, Shui and Yao etc. The ethnic minorities account for 67.08% of the total population. This Landscape Bridge City, with nearly 100 bridges of various architectural styles, straddles the Jianjiang River which embraces the whole city.

第二篇 国茶之都

“蟒山滴翠，剑水流芳”。有着“山水桥城 · 国茶之都”称号的都匀市，山清水秀，民族风情浓郁，是贵州省黔南布依族苗族自治州州府所在地，中国的优秀旅游城市，全球绿色城市。全市总面积 2274 平方千米，总人口 50 余万，有布依族、苗族、水族、瑶族等 33 个少数民族，少数民族占总人口的 67.08%。这座山水桥城，拥有近百座各种不同建筑风格的桥梁，横跨在环抱全城的剑江河上。

都匀石板街（作者：于洲桐）

经典 22

都匀石板古街

都匀石板古街始建于明洪武年间，路面以万余块青石铺砌而成，街道两旁的房屋皆以明清古典风格的青瓦、红墙、雕花门窗建造，街两头各立一对石狮，南面街口建有古式大门。这条街是明清时期贵州通往广西必经的古驿道。明崇祯十一年（1638 年），中国伟大的探险家、旅行家徐霞客即沿石板古街进入都匀府。

Duyun ancient slate street was built in Hongwu period of Ming Dynasty. The road surface was paved with 10001 bluestones. The houses on both sides of the street were built with gray tiles, red walls and carved wooden doors and windows, which were all classical Ming and Qing styles. A pair of stone lions presides over each end of the street, with an ancient gate built at the south end. This street was the only ancient post road connecting Guizhou and Guangxi in Ming and Qing Dynasties. In the 11th year of Chongzhen in the Ming Dynasty, Xu Xiake, a great Chinese explorer and traveler, entered Duyun along the ancient slate street.

经典 23

都匀东山晓日

都匀东山晓日（作者：于洲桐）

东山之巅有奎星阁，始建于明洪武二十七年(1394 年)。清嘉庆五年(1800 年)改名钟鼓楼。阁内置铜铸巨钟大鼓。每当清晨日出，自城中望去，山巅流光溢彩，楼阁若隐若现，仿佛琼楼仙阁，瑰丽无比。晨钟荡响，洪亮悠扬，飘越城区上空，声震数十里。明清时期称“东山晓日”，为“都匀八景”之冠。

There is Kuixing Pavilion on the top of Dongshan mountain, which was built in the 27th year of Hongwu (1394) in the Ming Dynasty. In the year of Gengshen(1800) in the Jiaqing period of the Qing Dynasty, it was changed into the bell and drum tower. Inside the pavilion, there are huge copper bell and drum. When the sun rises in the morning, from the view of the city, the top of the mountain is shining brightly, and the attic is looming, magnificent and incomparable, like the fairy pavilion. In the morning, the bell rings, loud and melodious, floating over the city and shaking for tens of miles.

经典 24

都匀文峰塔

都匀文峰塔原名文笔塔，始建于明万历年间，系五层木塔。因年久失修，几度重建。1983 年，都匀市人民政府拨款维修文峰塔，建成高 23 米、七层六面的实心石塔。“柳州八贤”的张翀和东林党领袖邹元标先后来到都匀，奠定了都匀的文教基础。文峰塔镇卫南天，直指文曲星，寓意文星下界。文峰塔建成后，都匀人才辈出。

Duyun Wenfeng tower, formerly known as Wenbi tower, was built in the Wanli period of the Ming Dynasty. It was a five story wooden tower, which has been rebuilt several times due to its disrepair. In 1983, Duyun Municipal People's Government allocated funds for the maintenance of Wenfeng tower, which is now 23 meters high and has seven floors and six sides. Zhang chong of “Liuzhou Eight Sages” and Zou Yuanbiao, leader of Donglin Party, came to Duyun successively, laying the cultural and educatonal foundation of Duyun. Wenfeng Tower, guarding the southern sky, points directly to Wenqu star, which means Wenxing(literary star) goes down to earth. After the completion of Wenfeng tower, Duyun’s talents are endless .

都匀文峰塔（**作者：于洲桐**）

经典25 都匀百子桥

都匀百子桥（作者：于洲桐）

百子桥建于清乾隆年间，为都匀水竹寨唐姓乡贤捐资修建，也称唐家桥，位于剑江河上。桥长140米，宽8米，高11.5米，结构为七孔石拱桥。桥上建楼廊，耸翠亭分立两头，石栏画栋，极富民族风格，是都匀桥的典范，为省级文物保护单位。传说当年唐氏年过半百，膝下无子，欲架桥以求继嗣。桥建成后果然得以应验，唐氏一年后喜得贵子。据《诗经·大雅·思齐》："大姒嗣徽音，则百斯男。"故名"百子桥"。

Baizi bridge, also known as Tangjia bridge, was built in Qianlong period of the Qing Dynasty. It was donated by the country worthy surnamed Tang in Shuizhu village of Duyun. It is located on Jianjiang River. The bridge is 140 meters long, 8 meters wide and 11.5 meters high. It is a seven hole stone arch bridge. The gallery was built on the bridge, and the towering green pavilion was divided into two ends. The whole bridge, with stone railings and painted buildings, is very rich in national style. It is a model of Duyun bridge and a provincial cultural relics protection unit. It is said that the man with the surname of Tang was over half a hundred years old and had no children. He wanted to build a bridge to obtain children. It did come true. One year after the completion of the bridge, he was happy to have his son. So people call this bridge "Baizi(hundreds of sons) bridge".

经典 26

都匀剑江河景区

都匀剑江河是都匀人民的母亲河，它波光粼粼，碧波荡漾，蜿蜒穿城而过，形成独特的“一江两城”和十里剑江看匀城美景之特色。剑江景区是都匀斗篷山—剑江风景名胜区的组成部分，是一个集自然山水、园林、桥梁、文物古迹、民族风情为一体的独具特色的风景区，其中沿河建造的元明时期至今的风格各异的桥梁有近百座。

Duyun Jianjiang River is the mother river of Duyun people. It is sparkling, rippling and winding through the city, forming a unique feature of one river two cities and ten miles Jianjiang River to see the beautiful scenery of Duyun city. Jianjiang River scenic area is a unique scenic area integrating natural landscapes, gardens, bridges, cultural relics, and national customs. Nearly 100 bridges with different styles have been built along the river since the Yuan and Ming dynasties.

都匀西山大桥（**作者：于洲桐**）

都匀南沙洲（**作者：于洲桐**）

经典 27

都匀南沙洲绿地公园

都匀南沙洲绿地公园位于市中心城区，剑江河环抱四周，被誉为都匀市的“肺叶”“城市会客厅” “都匀的绿岛” 。园内以黔南厚重的历史文化、秀美的自然生态以及独特的民族风情为主题，融入全州具有影响力的精品景点，打造成一个区域性的具有浓郁文化氛围的开放性公园。

Nanshazhou Green Park is located in the downtown area of Duyun City, surrounded by Jianjiang River. It is known as the “lung leaf”, “City reception hall” and “Green Island” of Duyun City. The theme of the park is the profound history and culture, beautiful natural ecology and unique national customs of Qiannan. It is integrated into the infuential boutique scenic spots of the whole prefecture to create a regional open park with rich cultural atmosphere.

经典 28

都匀青云湖森林公园

都匀青云湖森林公园 2005 年被国家林业局批准为国家级森林公园，是一个集森林景观、地貌景观、水体景观、动物景观、天象景观和独具民族特色的人文景观为一体的城市后花园，是紧挨城市中心区的绿色肺叶。

Qingyun Lake Forest Park was approved as a national forest park by the State Forestry Administration in 2005. It is a city back garden integrating forest landscape, geomorphic landscape, water landscape, animal landscape, celestial landscape and unique cultural landscape. It is a green lung leaf close to the city center.

都匀青云湖（作者：单继平）

都匀杉木湖（作者：单继平）

经典29 都匀杉木湖中央公园

杉木湖中央公园位于都匀市匀东镇，是一个集生态、文化、休闲、娱乐、健身、防灾功能为一体的综合性市政公共公园，总投资约3亿元。园中立有一“黔南星”雕塑，以纪念中国科学院国家天文台1997年9月26日发现的编号为第24956号小行星。

Shamu Lake Central Park is located in Yundong Town, Duyun city. It is a comprehensive municipal public park integrating ecology, culture, leisure, enterainment, fitness and disaster prevention functions, with a total investment of about 300 million yuan. There is a sculpture of "Qiannan star" in the park to commemorate the discovery of asteroid No. 24956 by the National Observatory of the Chinese Academy of Sciences on September 26, 1997.

都匀三江堰（作者：于洲桐）

经典 30

都匀三江堰湿地公园

三江堰位于都匀市茶园河、杨柳街河、摆楠河这三条河汇流的下游，剑江河的上游。三江堰湿地公园是一个多功能综合性湿地公园，建有水上游乐园、儿童游乐园、民族文化长廊、垂钓区、徒步健身区、商业街等 6 大功能区，已然成为剑江河上游的“净水器”。

Sanjiang Weir is located in the downstream of the confluence of the three rivers of Duyun city, i.e. Chayuan River, Yangliujie River and Bainan River, and the upstream of Jianjiang River. It is a multi-functional comprehensive Wetland Park with six functional areas, including water park, children's amusement park, national cultural corridor, fishing area, hiking fitness area and commercial area. It has become the "water purifier" in the upstream of jianjiang River.

经典31 都匀斗篷山景区

斗篷山景区位于都匀市西北部，是都匀斗篷山—剑江风景名胜区五大组成部分之一，距市区 22 千米，雄踞于苗岭山脉中段，其主峰海拔高度 1961 米，是国内距离城市最近的原始林区，原始森林覆盖率近 90%。其中，海拔 1800 米的高山台地上，有原始古林近百公顷，林木根部全都长在岩石缝隙之中，随处可见树抱石、石抱树、树搭桥的奇异景观。

Doupeng Mountain scenic spot is located in the northwest of Duyun City, 22 kilometers away from the urban area, occupying the middle of the Miaoling mountains. Its main peak is 1961 meters above sea level, which is the nearest primeval forest to the city in China, with a primeval forest coverage rate of nearly 90%. Among them, there are nearly one hundred hectares of primitive ancient forest on the high mountain platform with an altitude of 1800 meters. The roots of the trees are all growing in the rock cracks. There are strange landscapes of trees holding stones, stones holding trees and trees bridging.

都匀斗篷山（作者：于洲桐）

经典 32

都匀螺蛳壳风景区

螺蛳壳风景区位于市区西部约 20 千米处，其主峰螺蛳壳山为斗篷山的姊妹山，因山形酷似巨大的螺蛳而得名，总面积 45 平方千米，最高峰海拔 1738 米，为全市第二高峰。螺蛳壳山因其得天独厚的地理气候环境成为都匀毛尖茶的主要产区之一。

Luosike scenic spot is located in the west of the city, about 20 kilometers away. Its main peak, Luosike mountain, is the sister mountain of Doupeng mountain. It is named after the mountain shaped like a huge screw. With a total area of 45 square kilometers and the highest altitude of 1738 meters, it is the second highest peak in the city. Because of its unique geographical and climatic environment, Luosike mountain has become the main production area of Duyun Maojian tea.

都匀螺蛳壳（作者：于洲桐）

都匀茶文化博览园（作者：单继平）

经典 33

中国茶文化博览园

中国茶文化博览园位于都匀市匀东镇，占地 123 亩，重现了 1915 年巴拿马万国博览会上中国茶馆古建筑群的壮观景象，包括茶博园大门楼、八角塔、茶产品交易中心、茶文化体验中心、陈列馆。园区周边有茶文化广场、茶文化主题公园、百年毛尖古镇、饮食城等，现已成为国内乃至国际极具特色的茶文化博览园。

China Tea Expo Park is located in Yundong Town, Duyun city, covering an area of 123 mu. It reproduces the spectacular scene of the ancient Chinese teahouse complex at the 1915 Panama World Expo, including the gate building, octagonal tower, tea product trading center, tea culture experience center and exhibition hall. Around the park are tea culture square, tea culture theme park, century old Maojian ancient town, Catering City, etc. It now has become a unique tea culture Expo Park at home and abroad.

都匀毛尖小镇（作者：单继平）

经典 34

都匀毛尖小镇

都匀毛尖小镇，占地 900 亩，总投资 2 亿元。原为“三线建设”时期 883 厂旧址——曾经隐藏在大山深处的一所军工企业。进入城楼，林荫小道上铺满了青石板，一栋栋明清风格的小木楼依山而立，与山上的茶园相映衬，增添了小镇质朴沧桑的味道。这里注定是一个爬满故事的文艺小镇。以都匀毛尖为主题的大型电视连续剧《星火云雾街》即在这里拍摄完成。

Duyun Maojian town covers an area of 900 mu, with a total investment of 200 million yuan. It was the former site of 883 factory during the Third Line Construction period, and was once a military enterprise hidden in the deep mountains. When entering the ancient town, the path is covered with bluestone slabs. One after another, the small wooden buildings with Ming and Qing style stand on the mountain, which are set off with the tea garden on the mountain, adding the simple and vicissitudes of the town. It's destined to be a literary town full of stories. This is where the large TV series "Spark Yunwu Street" with the theme of Duyun Maojian was shot.

都匀螺蛳壳茶圣殿（**作者：单继平**）

经典 35

都匀螺蛳壳茶圣殿

螺蛳壳茶圣殿地处都匀毛尖茶核心产区、螺蛳壳风景区高腰平台，背靠都匀高寨水库，视野开阔，螺蛳壳旖旎风光和茶园的自然景色一览无余。

Duyun Luosike Tea Temple is located in the core production area of Duyun Maojian tea and the high waist platform of Maojian town scenic area in Duyun City. It is backed by Duyun Gaozhai reservoir and has a wide view. The beautiful scenery of Luosike and the natural scenery of tea garden can be seen at a glance.

经典 36

都匀秦汉影视城

秦汉影视城是集剧组拍摄一站式服务、弘扬体验汉文化为一体的影视体验基地，为西南地区最大的古代宫殿群影视城。

Qinhan film and television city is a film and television experience base that integrates one-stop shooting service of the drama group and promoting the experience of Han culture. It is the largest ancient palace group film and television city in Southwest China.

都匀秦汉影视城（作者：单继平）

都匀三线博物馆（作者：单继平）

经典37 都匀三线建设博物馆

二十世纪六十年代，党中央在中西部13省区部署了一场以战备为核心的大规模国防、科技、工业和交通基础设施建设，波澜壮阔的“三线建设”拉开帷幕。贵州都匀成为西南“三线建设”的电子军工基地。为重现“三线建设”时期的流芳岁月，都匀市在东方机床厂工业遗址上沿脉建成“都匀三线建设博物馆”，成为集大“三线”、大历史、大未来为一体的复合型文化广场。这是对尘封记忆的唤醒，更是对光辉历史的弘扬。

In the 1960s, the CPC Central Committee deployed a large-scale movment of national defense, science and technology, industry and transportation infrastructure construction with war preparation as the core in 13 provinces and regions in the central and western regions. The magnificent "Third Line Construction" started. Duyun, Guizhou Province, had become the electronic military industry base of "Third Line Construction" in Southwest China. In order to recreate the most uncommon years of the "Third Line Construction", Duyun municipal government has built a "Third Line Construction Museum" along the vein on the industrial site of Duyun Dongfang machine tool plant, which has become a composite Cultural Square integrating the great "Third Line", the great history and the great future. This is the wake-up of the dust laden memory and also the promotion of the glorious history.

都匀绿博园（**作者：单继平**）

经典 38
第四届中国绿化博览园

第四届中国绿化博览会在都匀举办。这是我国国土绿化领域组织层次最高、展示内容最多、影响力最大的国家级综合性博览会，也是第一次在中国西部少数民族地区山地城市举办。都匀绿博园总面积 1959 公顷，森林覆盖率 69%。整个园区山、水、林、田、湖、草各种生态元素齐聚，一岛两湖，四处滨湖生态湿地，七座桥，沿湖多个亭、廊、榭、码头等滨水景观，有鸟栖湿地、行舟飞幕、七彩梯田、彩叶花谷等环湖十景，及十二条花色景观大道。整个园区月月有风景，季季有花香。

The 4th China Green Expo will be held in Duyun. This is the national level comprehensive exposition with the highest level of organization, the most display content and the most influence in the field of land greening in China, and it is also the first time held in the mountainous cities in the western minority areas of China. The total area of Duyun Green Expo Park is 1959 hectares, and the forest coverage rate is 69%. All kinds of ecological elements gather in the park, and there are sceneries in month and flowers fragrance in season.

Chapter III
Famous for Thousands of Years

Duyun Maojian tea, the king of auspicious grass, has the sweet dew of heaven and earth, the essence of humanities, the spirit of the universe and the soul of the nation. Its culture is attached to the historical ark of Chinese civilization and drifted here along the long history river. From thin to thick, and naive to mature, after countless generations of accumulation, it has finally become a cultural carrier with distinctive style, bearing the deep feelings of the Chinese nation.

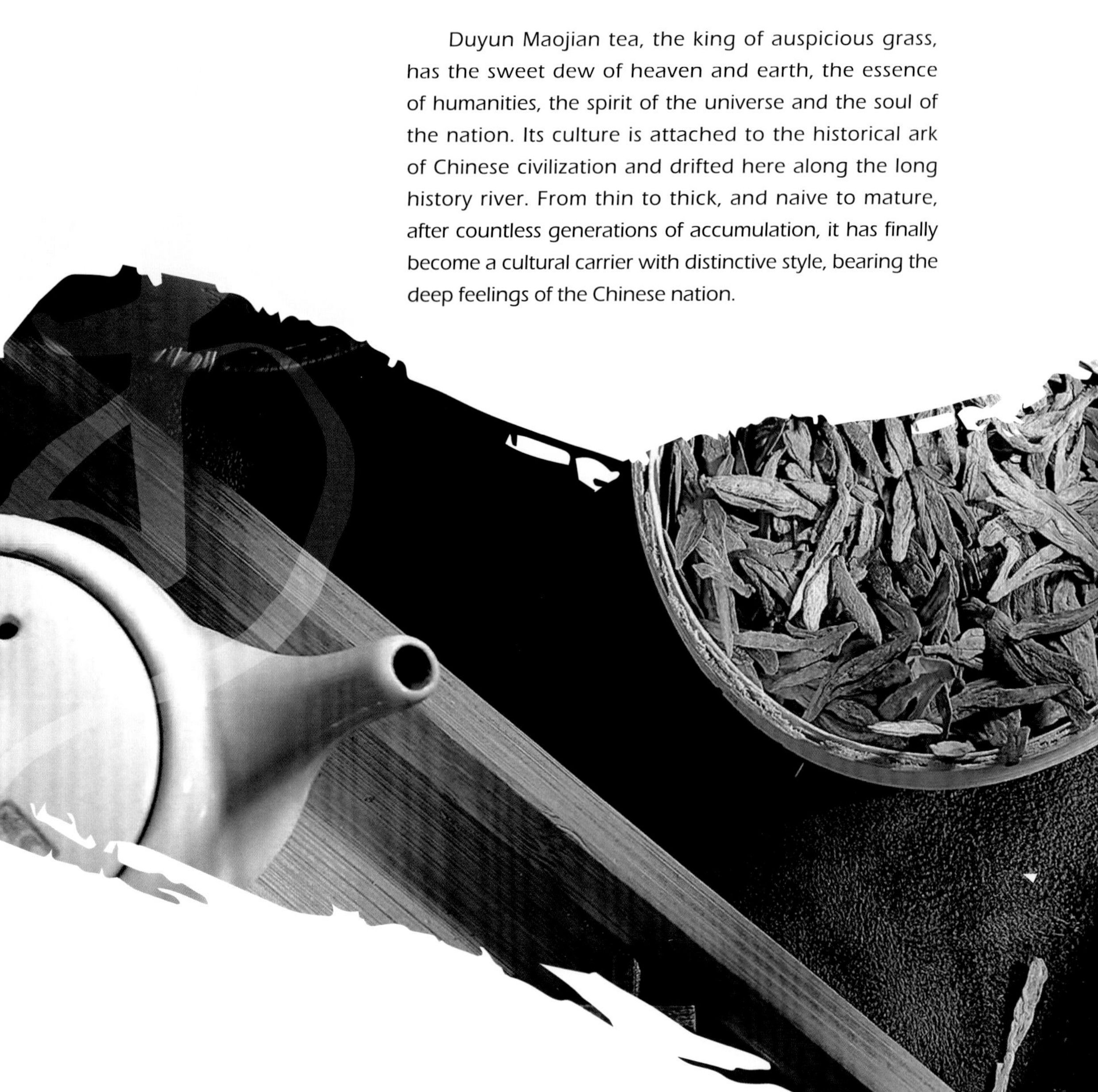

第三篇 名垂千古

“瑞草魁”都匀毛尖，得天地之甘露，咀人文之精华，涵乾坤之灵性，育民族之魂魄。其文化依附于华夏文明的历史方舟，沿着漫漫历史长河漂流而来，从单薄走向厚重，从幼稚走向成熟，经过无数代的积淀，终于成为一种彰显风格的文化载体，承载着中华民族的深深情结。

毛尖与名人

Tea is the symbol of Chinese civilization. Probably because China is the "mother country of tea", the leaders of modern China seem to have inextricably linked with tea, and left many pleasant talks or stories related to tea. At the same time, no tea, no literati. A cup of bitter tea and a literati have achieved a kind of tacit communication between them in silent eyes. The literati borrow tea to cultivate quality and moral character, and borrow tea to enlighten. Tea borrows the literati to sublimate.

茶是中国文明的象征。大抵因为中国是"茶之母国"的关系，近现代中国的名人，似乎都与茶有着千丝万缕的机缘，并留下了不少与茶有关的美谈或故事。同时，无茶不文人。一杯清茶，在与文人的默默对视中，相互间完成了一种相濡相融的默契。文人借茶以养素，借茶以修身，借茶而开悟。茶借文人以升华。

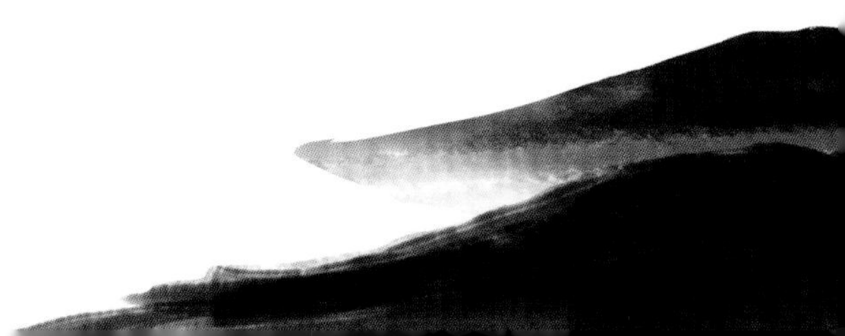

经典 39

赐名都匀毛尖

清供图（作者：王林）

有一位东方伟人曾盛赞黔南茶：“高山云雾出好茶哟！”并称此茶为毛尖，都匀毛尖茶由此得名。

A great man in the East once praised Qiannan tea: "its a good tea comes out of high mountain clouds!" and called this tea Maojian, which was named Duyun Maojian tea .

经典 40

多次获得赞誉的都匀毛尖

有领导发出“对于都匀毛尖，希望你们把品牌打出去”的重要指示。强调中国是茶的故乡，从古代丝绸之路、茶马古道、茶船古道，到今天丝绸之路经济带、21 世纪海上丝绸之路，茶穿越历史、跨越国界，深受世界各国人民喜爱。从文化到外交，一杯茶，正是中国“和而不同”理念的彰显和展现。

One leader has issued an important instruction to “for Duyun Maojian, I hope you can promote Duyun Maojian’s brand out." He emphasized that China is the hometown of tea. From the ancient Silk Road, the ancient tea horse road, the ancient tea boat road, to today's Silk Road Economic Belt and the 21st century Maritime Silk Road, tea has crossed history and borders, and is deeply loved by people all of the world. From culture to diplomacy, a cup of tea is the manifestation of China's concept of "homony but difference".

清供图（作者：周艺）

黄庭坚（作者：王殿维 临）

经典 41

黄庭坚《阮郎归》赞毛尖

北宋著名文学家、书法家黄庭坚，文学上与苏轼齐名，世称“苏黄”。他一生爱茶，写下了不少赞茶诗词，其中《阮郎归》赞美黔南茶（即都匀毛尖）：“黔中桃李可寻芳，摘茶人自忙。月团犀胯斗圆方，研膏入焙香。青箬裹，绛纱囊，品高闻外江。酒阑传碗舞红裳，都濡春味长。”

Huang Tingjian was a famous litterateur and calligrapher in the Northern Song Dynasty. He was as famous as Sushi in literature and was known as “Suhuang” in the world. He loved tea all his life and wrote many poems praising tea, among which Ruan Langgui praised Qiannan tea (that is, Duyun Maojian tea).

经典 42

“明朝柳州八贤”张翀的都匀茶联

明代“柳州八贤”之一的张翀任刑部主事时弹劾严嵩，被加害入狱，后被朝廷谪戍都匀。一个春天，被贬谪都匀 8 年的张翀出游山水，饱尝都匀毛尖后回到寓所，提笔写下一幅茶联：“云镇山头，远看青云密布；茶香蝶舞，似如翠竹苍松。”

When Zhang Chong, one of the “Eight Sages of Liuzhou” in the Ming Dynasty, was the head of the Ministry of Punishment, he impeached Yan Song. He was persecuted and jailed, and then was sent into exile in Duyun by the imperial court. One spring, Zhang Chong, who has been relegated to Duyun for 8 years, went out to visit the mountains and rivers. After he tasted Duyun Maojian, he returned to his apartment and wrote a tea couplet praising Duyun Maojian tea.

张翀（作者：王长存）

经典 43

“西南巨儒”莫友芝与采茶诗

莫友芝乃都匀府独山兔场人，清嘉庆十六年（1811 年）生，与郑珍并称为“西南两大儒”，两人主编的《遵义府志》，被梁启超誉为“天下府志第一”。莫友芝的书法被列入我国禁止外流的书画作品目录，属于国宝级作品。莫友芝生平嗜茶，在《遵义府志》中，他收录了一首流传于独山、都匀、福泉一带的花灯《采茶调》中的《十二月采茶歌》，歌云：“三月采茶茶叶清，茶树脚下等莺莺；二月采茶茶花开，借问情侬几时来……”

Mo Youzhi, a native of Dushan Tuchang in Duyun, was born in the 16th year of Jiaqing in the Qing Dynasty. He and Zheng Zhen were called “two great Confucians of the southwest”. Zunyi Official Records edited by the two men was praised by Liang Qichao as “the first Official Records in the world”. Mo Youzhi’s calligraphy has been listed in the catalogue of calligraphy and painting works prohibited from outflow in China, which belongs to the national treasure level works. Mo Youzhi was fond of tea in his life. In Zunyi Official Records, he collected a Tea Picking Song of Twelve Monthes in the festive lantern opera “Tea Picking Tune”, which was popular in Dushan, Duyun and Fuquan.

莫友芝（作者：王长存）

纪晓岚（作者：王殿维 临）

经典 44

“学宗汉儒”纪晓岚与都匀毛尖

清代政治家、文学家纪晓岚于乾隆三十三年（1768 年）被授予贵州都匀知府。因乾隆皇帝惜才，又将其留在了身边。纪晓岚虽未到都匀履职，但心中已有了都匀。深知乾隆嗜茶，便差人到贵定云雾弄来“鸟王茶”（即现在的“云雾茶”，属都匀毛尖品牌。）进贡乾隆。乾隆饮罢盛赞，遂欣然提笔改名为“仰望茶”。在纪晓岚的努力下，朝廷调拨白银，改造、扶持当地茶园，并两立“贡茶碑”。

Ji Xiaolan, a politician and litterateur in the Qing Dynasty, was awarded the prefect of Duyun in Guizhou Province in the 33rd year of Qianlong. Because Emperor Qianlong cherished his talent and left him by his side. Although Ji Xiaolan did not arrive at Duyun to perform the duties, he had Duyun in mind. Knowing that Qianlong loved tea, he sent people to Guiding Yunwu to get “bird king tea” (a brand of Duyun Maojian tea) to pay tribute to Qianlong. After drinking and praising, Qianlong gladly changed the name of the tea to “looking up tea”. With the efforts of Ji Xiaolan, the imperial court allocated silver to transform and support the local tea garden, and set up “tribute tea stele” twice.

贵州巡抚林绍年进贡毛尖

林绍年（作者：王殿维 临）

清同治十三年（1874 年），进士林绍年——曾任贵州巡抚两年，每年特意精选精装都匀毛尖云雾茶——专程进京向光绪皇帝和慈禧太后进贡。《清宫密档》记载："贵定县芽茶，贡皇上一匣，贡老佛爷一匣。奴才贵州巡抚林绍年叩首。"

In the 13th year of Tongzhi in the Qing Dynasty, Lin Shaonian, a Jinshi, served as governor of Guizhou for two years. Every year, he specially selected Yunwu tea(a brand of Duyun Maojian tea) and went to Beijing to pay tribute to Emperor Guangxu and Empress Dowager Cixi. "Secret Records of the Qing Palace" records: "bud tea in Guiding County, a box for the emperor, a box for the old Buddha. My servant, Lin Shaonian, governor of Guizhou Province, kowtowed."

经典 46

“茶树栽培学科奠基人”庄晚芳咏都匀毛尖

中国茶学家、茶学教育家、茶叶栽培专家庄晚芳先生，是中国茶树栽培学科的奠基人之一，他毕生从事茶学教育与科学研究，培养了大批茶学人才，在茶树生物学特性和根系研究方面取得了成果。二十世纪六十年代末，都匀毛尖茶从色、香、味、形、效方面均有突破，为此庄先生给予了很高评价，并题诗赞道：“雪芽芬芳都匀生，不亚龙井碧螺春。饮罢浮花清鲜味，心旷神怡攻关灵！”

Mr. Zhuang Wanfang, a Chinese tea scientist, tea educator and tea cultivation expert, is one of the founders of Chinese tea cultivation discipline. He has been engaged in tea education and scientific research all his life, and has trained a large number of tea science talents. He made achievements in the research of tea biological characteristics and root system.

庄晚芳（作者：王长存）

经典 47

『开创媒介宣传茶叶先河』的张大为题诗都匀毛尖

张大为（作者：王长存）

张大为曾任中国茶叶学会理事、中国国际茶文化研究会常务理事等职务，是中国茶叶文化宣传的“先驱者”。退休后热衷于中国茶文化宣传和研究，1989 年创建了国内首家茶道馆——北京茶道馆。首创了茉莉花茶茶艺表演，编著有《中国茶馆》《茶艺师培训教材》等，极力推崇“俭、美、和、静”茶德。在品评都匀毛尖后，题诗道：“不是碧螺，胜似碧螺。香高味醇，别具一格。”

Zhang Dawei once served as the director of China Tea Society and executive director of China Internatioanl Tea Culture Research Association, and was the “pioneer” of Chinese tea culture publicity. After retirement, he was keen on the promotion and research of Chineses tea culture, and in 1989, he established the first teahouse in China: Beijing Teahouse. He initiated the performance of jasmine tea art, and compiled books such as Chinese Tea House and Tea Art Specialist Training Materials, and highly praised the tea virtues of “thrift, beauty, harmony and quietness”.He once wrote poems praising Duyun Maojian tea.

经典 48

“一代佛学大师”赵朴初与黔南“佛茶”

赵朴初（作者：王长存）

贵定县的阳宝山石塔林是国内规模最大的石刻和尚坟塔林，可与中原少林寺砖塔林媲美，人称“北有少林砖塔，南有阳宝石塔”。阳宝山上所产名茶均系开山白云祖师、宝华上人、然薄大师等历代高僧亲手培植。1997 年，时任中国佛教协会会长赵朴初老先生在亲自品尝贵定云雾春茶后，兴之所至，感慨万千，欣然挥毫，题写了“佛茶”二字。由此，黔南茶增添了历史贡茶与佛教文化共融一体之神秘色彩，也是唯一真正得到佛教界权威人士认可并题名的“佛茶”。

Yangbao Mount Stone Pagoda Forest in Guiding county is the largest stone carving Monk Tomb Pagoda Forest in China. It can be compared with the brick Pagoda Forest of Shaolin Temple in the Central Plains. It is called “there is Shaolin brick pagoda in the north and Yangbao stone pagoda in the south”. The famous tea produced in Yangbao Mount is cultivated by the great monks of the past dynasties, such as the founder of Baiyun, the master of Baohua, and the master of Ranbo. In 1997, Mr. Zhao Puchu, then president of the Chinese Buddhist Association, was very excited after tasting the spring tea of Guiding Yunwu, and wrote the word “Buddha tea” with great emotion. As a result, Qiannan tea has added the mysterious color of the integration of historical tribute tea and Buddhist culture, and it is also the only “Buddha tea” that is really recognized and named by Buddhist authorities.

毛尖茶史

"Tea fragrance has a long history of thousands of years, and the color of tea does not reduce the eternal feeling." Duyun is the hometown of tea. The tea culture is rooted in Chinese culture. Thousands of years of gestation and development have accumulated the profound history of Duyun Maojian tea, and cleaned up the immortal spirit of all nationalities in this land.

"茶香幽远千年史，茗色不减万古情。"

都匀是茶的故乡，茶文化根植于中华文化。千载的孕育和发展，积蓄了都匀毛尖茶历史的浑厚，涤荡着这块土地上各民族不朽的性灵。

经典 49

千年贡茶史

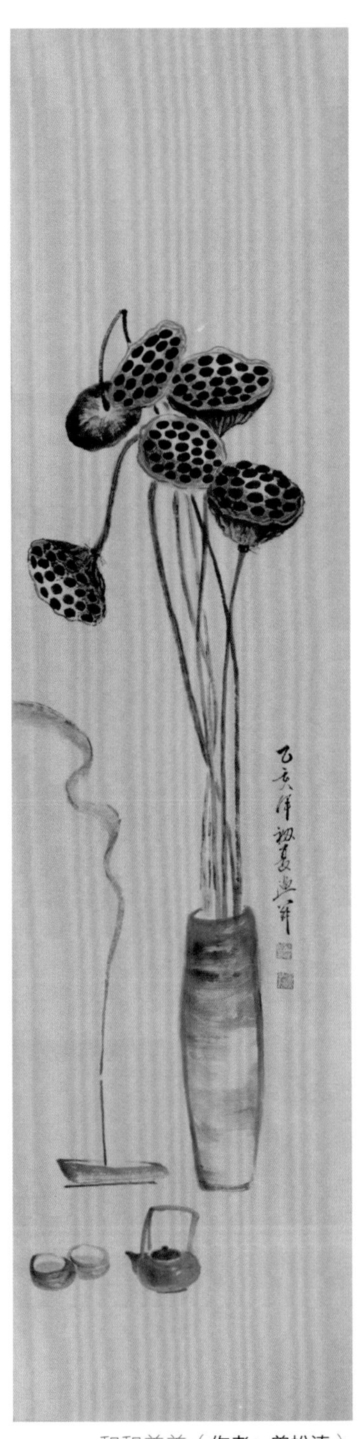

和和美美（作者：曾松涛）

在中国古代茶叶史上，黔南绿茶（都匀毛尖前身）是拥有贡茶原始“证据”最多的茶叶。进入朝廷的历史可以追溯到3000多年前的商周时代，是部落敬献王室的礼品。当巴人“前歌后舞”列队将茶叶贡献给周王朝时，黔南茶就开始了挺进宫廷的步伐，并在唐代达到鼎盛。

In the history of ancient Chinese tea, Qiannan green tea (the predecessor of Duyun Maojian) has the most original "evidence" of tribute tea. The history of entering the court can be traced back to the Shang and Zhou dynasties more than 3000 years ago. It is a gift offered by the tribe to the royal family. When the Ba people "sang before and danced after" lined up to contribute tea to the Zhou Dynasty,Qiannan tea began to march into the imperial court, and reached its peak in the Tang Dynasty.

品茶论道（作者：傅宝）

经典 50

陆羽赞毛尖

周朝，都匀、福泉等地开始人工栽培茶树，茶叶不但进入市场进行交易，且已成贡品。唐朝“茶圣”陆羽曾评价黔南茶叶（都匀毛尖前身）：“往往得之，其味极佳。”说明黔南茶品质上乘，且早已名声远扬。

In the Zhou Dynasty, tea trees were cultivated in Duyun, Fuquan and other places. Tea not only entered the market for trading, but also became a tribute. Lu Yu, the "tea sage" of the Tang Dynasty, once commented that Qiannan tea(the predecessor of Duyun Maojian) "often get it, its taste is excellent. " It shows that Qiannan tea is of high quality and has long been famous.

品茶图（作者：李彦岭）

经典51

朱元璋『罢造龙团』

明洪武皇帝朱元璋深知，制作精致龙凤团饼贡茶，需耗费大量人力物力。当他发现都匀进贡的散绿茶能直接沏泡，可取多可撮少，十分方便时，遂于明洪武二十四年（1391 年）颁布了一道茶叶历史上的著名诏书："罢造龙团，惟采芽茶以进。"终结了团茶独领风骚的生产历史。

Zhu Yuanzhang, Emperor Hongwu of the Ming Dynasty, knew that it took a lot of manpower and material resources to make the tribute tea with exquisite dragon and phoenix pancakes. When he found that the scattered green tea from Duyun could be brewed directly, which could be more or less, and was very convenient, he issued a famous imperial edict in the history of tea in the 24th year (1391) of Hongwu in Ming Dynasty: "stop making the dragon pancake of tea, and only take bud tea to pay for tribute." The production history of pancake of tea for hundreds of years ended.

经典52

崇祯皇帝赐名『鱼钩茶』

清供图（作者：王林）

明崇祯皇帝朱由检对时任辽东巡抚丘禾嘉（贵定人）所贡黔南茶（都匀毛尖前身）高度赞赏：“卿所贡之茶，历朝有名。”他根据“生时为枪，熟时为钩”的特性，赐名黔南茶为“鱼钩茶”。从此，黔南茶有了品茶、赏茶的标准。

Zhu Youjian, Emperor Chongzhen of the Ming Dynasty，highly praised Qiannan tea (the predecessor of Duyun Maojian) which was paid by Qiu Hejia(Guiding people), the governor of Liaodong at that time: “the tea you paid is famous in all Dynasties.” He named Qiannan tea “fish hook tea” according to the characteristics of “spear when born and hook when ripe”. Since then, Qiannan tea has the standard of tasting and appreciating tea.

其乐无穷（作者：傅宝）

经典 53

都匀西岳茶园

清乾隆前期，官方在今天的都匀市团山一带建西岳庙镇守茶园，还设有专理茶事的机构，由都匀知府宋文型兼理，以期“上裕国课，下佐工商”。据载，都匀西岳茶园是当时全国的第一个茶叶“国企”。

In the early Qianlong period of the Qing Dynasty, the government built Xiyue temple in Tuanshan area of today's Duyun City to guard the tea garden. There is also a special organization for tea affairs, which is jointly managed by the governor of Duyun, Song Wenxing, in order to "increase the state tax and assist the industry and commerce". According to the reports, Duyun Xiyue tea garden was the first tea "state-owned enterprise" in China at that time.

经典54

黔南『贡茶碑』

清品图（作者：王林）

清朝，在贵定云雾一带，由官方拨给白银四百二十两，扶持云雾鸟王十八寨，改造衰老低产的贡茶园，并树立“贡茶碑”界定贡茶产地区域，明确每年贡茶数量。据考证，此“贡茶碑”是目前国内唯一被发现并保存完好的有关茶叶的碑石，同时也是国内最早的茶叶生产“地理标志”。

In the Qing Dynasty, in the Yunwu area of Guiding, the government allocated 420 liang of silver to support the 18th village of Yunwu bird king to transform the old and low yield tribute tea garden, and set up a "Tribute Tea Monument" to define the origin area of tribute tea, and define the number of tribute tea each year. According to research, this "Tribute Tea Monument" is the only stone tablet related to tea found and well preserved in China, and also the earliest "geographical indication" of tea production in China.

赏石悟道图（作者：马纪奎）

经典 55
“北有茅台，南有毛尖”

1915 年，在美国旧金山举办的“巴拿马万国博览会”上，都匀选送参展的茶叶获得优奖。这是都匀毛尖茶获得的第一个国际性荣誉，世人将其与同时获得优奖的贵州茅台酒并称为“北有茅台，南有毛尖”。

In 1915, at the Panama World Expo held in San Francisco, the United States, the tea selected by Duyun won the excellent prize. This was the first international honor that Qiannan tea (the predecessor of Dujuan Maojian) had won. People call it “Maotai in the north and Maojian in the south” together with Guizhou Maotai liquor, which also won the excellent prize in the Expo.

经典 56 都匀『无垢茶』

茶趣（作者：森然）

都匀斗篷山清塘村的无垢茶，是都匀毛尖的茶中之王。这种栽培在房前屋后岩缝间的半乔木状茶树，茶叶中茶碱含量少，冲泡后茶汤如玛瑙般黄里透亮，清淡幽香，静置茶具中，无论时间多长，其表面都不会形成油状薄膜，茶具内也不会积淀咖啡色茶垢，所以本地人称“无锈茶”“无垢茶”，也称“清塘茶”。茶客们更称其为“能喝的液体玛瑙”，产量极少。

The dirt free tea in Qingtang village, Doupeng Mountain, Duyun, is the king of Duyun Maojian tea. This kind of semi arbor tea tree is cultivated in the front and back of the house and between the rock seams. The content of theophylline in the tea is low. After brewing, the tea soup is as bright as agate yellow, light and fragrant. No matter how long it takes, no oil film will form on the surface of the tea and no coffee dirt will accumulate in the tea set. So local people call it “rust free tea”, “dirt free tea”, also known as “Qingtang tea”. The tea customers call it “the liquid agate that can drink” even more, and the output is very small.

经典 57

美国总统富兰克林·罗斯福偏爱独山高寨茶

据《独山县志》记载，1945 年抗战胜利后，为取得美国经济支持，宋美龄率团赴美访问。在准备礼品时，宋美龄了解到美国总统富兰克林·罗斯福偏爱独山高寨茶（属都匀毛尖），便亲自授意贵州购买了数十斤该茶作为国礼赠送给美国总统罗斯福。

According to the records of Dushan County Annals, after the victory of the Anti Japanese war in 1945, Song Meiling led a delegation to visit the United States in order to obtain economic support from the United States. When preparing the gift, Song Meiling learned that President Franklin Delano Roosevelt preferred Dushan Gaozhai tea (belonging to Duyun Maojian brand), so she personally instructed Guizhou to buy dozens of Jin of the tea as a national gift to President Roosevelt.

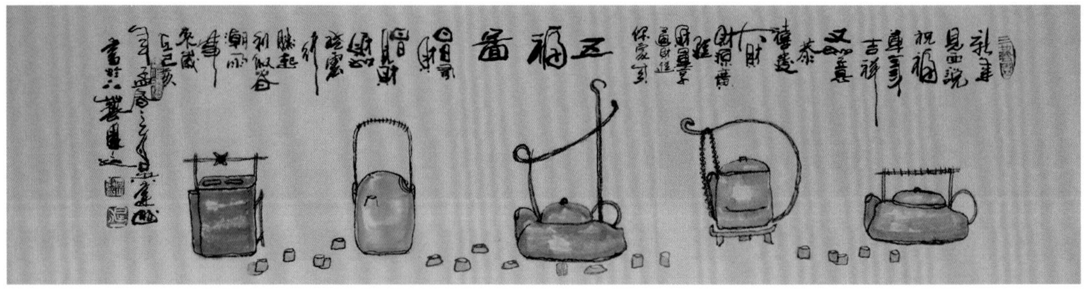

五壶图（作者：吴建国）

经典58
飞机换毛尖

书案小景（作者：王林）

日本前首相田中角荣，1972年访问中国时曾向周总理提出一个要求：用一架直升飞机换25公斤都匀毛尖茶。周总理听后笑了起来，幽默地回答："那我们的茶叶也太贵了！"因当时都匀毛尖茶年产量也不过300多斤，又时值9月，此事未成。不管此说是真是假，都匀毛尖的名气之大，品质之好，身价之高，可见一斑。

Former Japanese Prime Minister Kazuo Tanaka, when he visited China in 1972, made a request to Premier Zhou: use a helicopter to exchange 25kg Duyun Maojian tea. Premier Zhou laughed and replied humorously: then our tea is too expensive! At that time, the annual output of Duyun Maojian tea was only more than 150 kilograms, and it was in September, which was not completed. Whether it's true or not, it can be seen that Duyun Maojian tea is famous for its good quality and high value.

毛尖传说

Duyun Maojian is a long and profound historical ballad. She sings the glory and brilliance of Qiannan tea, which is rarely known by people, and also depicts her humanistic character full of wisdom and smart spirit. Thousands of years of historical immersion and cultural embellishment have made her into the deep layer of historical culture.

都匀毛尖是一首绵延悠长而韵味深厚的历史歌谣，唱出了黔南茶鲜为人知的荣耀与辉煌，也刻画出了它充满睿智、灵秀之气的人文品格，数千年的历史浸淫、文化点染，使得它已经进入历史文化的深层。

经典 59

都匀毛尖与蛮王

毛尖与蛮王（作者：王殿维 临）

相传在很早以前，都匀府地有个蛮王，他有九个儿子和九十个女儿。蛮王老了，有一次病得很重，他的那些儿女们就出去找药。九个儿子找回来九种药，老蛮王服用后都没效果。女儿们找回来的却都是云雾山上绿仙雀给的茶叶。老蛮王喝了，病很快就好了。于是希望她们能去找些茶种回来栽种。姑娘们就一起去寻找绿仙雀要茶种。经过一番艰苦努力，绿仙雀终于给了她们一些茶种，并不时喊着“毛尖、毛尖”。姑娘们将茶种带回去种植，经过精心栽培，很快长成了一片片茂密的茶园。蛮王有了茶园，国家也变得国泰民安。这个茶就是都匀毛尖茶。

It is said that a long time ago, there was a king in Duyun district. He had nine sons and ninety daughters. The king was old. Once he was very ill. His children went out to find drugs for him. Nine kinds of drugs were found by his nine sons, but they didn't work after old king took them. The girls found tea from the Green Fairy Bird on the cloud mountain. Old king soon recovered after drinking the tea, so he hoped his daughters could find some tea seeds to plant. The girls went to look for the Green Fairy Bird for tea. After a lot of hard work, the Green Fairy Bird finally gave them some kinds of tea seeds, sometimes shouting "Maojian, Maojian". The girls took the tea seeds back to plant. After careful cultivation, it soon grew into a dense tea garden. With the tea garden, the country became prosperous and the people lived in peace. This tea is Duyun Maojian tea.

毛尖与蛮王（作者：王殿维 临）

据《都匀府志》记载，都匀毛尖明初为上贡茶。传说明洪武年间，一支官兵驻扎都匀，因水土不服，很多士兵病倒了。当地一位布依族老人知道后，主动带上一些盐、茶、米、豆，煮汤给官兵喝，士兵全部治愈。后有一位将领偷带了一包毛尖回京城，让皇帝品尝，皇帝品后大喜，即令每年派专人来都匀收缴都匀毛尖作贡茶。

According to the Records of Duyun Prefecture: it was tribute tea in the early Ming Dynasty. It is said that during the Hongwu years of the Ming Dynasty, an army was stationed in Duyun, and many officers and soldiers fell ill because of acclimatization. After a local old man of Buyi nationality knew it, he took some salt, tea, rice and beans with him and cooked soup for the officers and soldiers to drink. All of them be cured. Later, a general secretly brought a bag of Maojian tea back to the capital for the emperor to taste. The emperor was very happy after tasting, so every year he sent a special person to collect Duyun Maojian as tribute tea.

经典61

都匀毛尖与神仙

毛尖与神仙（作者：王殿维）

很久以前，中国西南部曾是一片大海，大海中零星散布着一些岛屿。其中一座岛上的居民种有茶叶，形似鱼钩。村民长期喝这种茶，体健少病，村落兴旺繁荣。一天，有位名叫“都匀”的神仙路经这片岛屿，一股清香扑面而来，禁不住上岸一探究竟。村民们以茶相待。神仙饮罢，神清气爽，满口生津，颔首赞叹：“好香的毛尖茶。”于是脱下斗篷，放下宝剑，撂下金钵，飘然而去。受仙人点化，这座岛屿次日逐渐升高，变成了一座城市（都匀市），斗篷化作斗篷山，宝剑化作剑江河，金钵变为高寨水库，海里的贝螺则化成如今的螺蛳壳山。

Once upon a time, Southwest China was a sea with scattered islands. Among them, the islanders in one of the islands has planted tea, which looks like fish hook. The villagers drink the tea for a long time, and they are seldom sick. The village is prosperous. One day, an immortal named "Duyun" passed by this island. A fresh fragrance came, and he couldn't stand to go ashore to find out. The villagers treated him with tea. After drinking it, the immortal was refreshed and full of saliva. He nodded and exclaimed, "how fragrant Maojian tea is!". So he took off his cape, puting down his sword and golden bowl, and flew away. Enlightened by the immortal, the island gradually rose the next day, becoming a city (Duyun City). The cape became the Doupeng Mountain, the sword became the Jianjiang River, the golden bowl became the Gaozhai Reservoir, and the conch in the sea became the Luosike Mountain.

Chapter IV
Brand Creation

Sip a cup of clear tea to clean the dust belly. In the face of a cup of Maojian tea with such a heavy load, on the basis of inheriting the ancinet "tea painting", the famous artists have made great efforts to explore the aesthetic roots of tea painting art, and deeply explore the broad connotation of tea culture, so as to live up to the moistening of a cup of green tea.

第四篇 品牌缔造

一瓯清茗，轻啜入口，涤荡尘腹。面对承载着如此重负的毛尖茶，名家们在继承古代“茶事画”的基础上，借古鉴今，努力探索茶画艺术的美学根源，深入挖掘茶文化的博大内蕴，以此完成一次画笔与宣纸的惊艳对白，从而不负一杯清茗的润泽。

毛尖醉美

Tea is also intoxicating without wine. Four seasons rounding, clouds gathering and scattering, the stream in mountains running, Duyun Maojian tea, which is co-existing with karst spirit in South China, is permeated with the thousand-year old fragrance of ancient tea culture through the distant historical sunshine and the light across time and space.

茶亦醉人何须酒。四季轮回，云雾聚散，山溪奔流，和中国南方喀斯特灵气共生共荣的都匀毛尖，透过遥远的历史阳光，在跨越时空的光芒下，如绿云一般洋溢着古茶文化的千年陈香。

心上毛尖

心上毛尖（作者：舒肖）

黔南是茶树的原产地之一，孕育了赫赫有名的都匀毛尖。都匀毛尖的名气，源于她的品质，而她的品质取决于良好的种质基因，良好的种质基因又得益于黔南特有的自然条件。“要喝没有污染的茶就到贵州来”，这是贵州省委书记孙志刚向全世界发出的底气十足的邀请。

Qiannan is one of the origins of the tea trees, which breeds the famous Duyun Maojian. Duyun Maojian's fame comes from her quality, which depends on good germplasm genes, and good germplasm genes benefit from the unique natural conditions of Qiannan.

江边品茗图（作者：陈鸿铭）

据权威资料《黔南茶树种质资源》，黔南已调查收集到416份古茶树资源样本。其中，“都匀团山种”于明朝初年开始被规模种植，并建成皇家茶园；“龙里云台种”于明朝后期开始被规模种植，并建成寺庙茶园；“贵定鸟王种”于清朝康熙年间开始被规模种植，乾隆年间成为贡茶，同期“清塘种”开始被成片种植。

According to the authoritative data “Germplasm Resources of Qiannan Tea”, 416 samples of ancient tea resources have been collected in Qiannan. Among them, Duyun Tuanshan species began to be planted on a large scale and built into a royal tea garden in the early Ming Dynasty; Longli Yuntai species began to be planted on a large scale and built into a temple tea garden in the late Ming Dynasty; Guiding bird king species started to be planted on a large scale in the Kangxi year of the Qing Dynasty, became a tribute tea in the Qianlong year, and Qingtang species began to be planted on a large scale in the same period.

我要都匀
毛尖
她是最美
妙的東方
神葉

林志玲（作者：舒肖）

经典 64 领袖级好茶

在中国十大名茶中，林志玲是都匀毛尖的唯一代言人。2012年贵州国际绿茶博览会上，贵天下茶业公司花费880万元聘请林志玲代言都匀毛尖茶。沏茶、举杯的一瞬间，林志玲创下了中国茶史上第一位绝色佳人为茶代言的纪录，一笑一颦，生动地展现了宋代词人苏轼“欲把西湖比西子，浓妆淡抹总相宜”的意韵。

Among the top famous teas in China, Lin Zhiling was only the spokesperson for Duyun Maojian. At the 2012 Guizhou International Green Tea Expo, Lin Zhiling was hired to represent Duyun Maojian Tea for ¥8.8 million yuan. At the moment of tea making and cup raising, Lin Zhiling set a record of the first matchless beauty in the history of Chinese tea to speak for tea.

以茶兴业（作者：尚秋香）

经典 65

论道都匀毛尖

龙永图曾在《论道都匀毛尖》中语惊四座：“你可随处建厂制造出原子弹，离开了都匀毛尖茶的原产地，你就合成不出一片毛尖茶。”是的，都匀毛尖千年传承，历久弥香，雄伟厚重。都匀毛尖顺应时代，持续发展，春山可望。

Long Yongtu once said on “Duyun Maojian Forum”, capturing the attention of all present : “You can build factories everywhere to make atomic bombs. Without the origin of Duyun Maojian tea, you can’t synthesize a piece of Maojian tea.” Yes, Duyun Maojian has been passed down for thousands of years. It’s fragrant and majestic. Duyun Maojian complies with the times and develops continuously. It’s very hopeful.

都匀毛尖茶品牌简介（作者：李传平）

经典66

都匀毛尖茶2018新标准

都匀毛尖茶以采自贵州省黔南布依族苗族自治州境内的中小叶茶树群体种或适宜的茶树良种的幼嫩芽叶为原料，按 DB52/T 433 规定加工而成，是具有特定品质特征的卷曲形绿茶。按照原料标准，结合加工工艺和生产实际，都匀毛尖茶分为五个等级：尊品、珍品、特级、一级、二级。

Duyun Maojian tea is a curly green tea with specific quality characteristics, which is made from the young buds and leaves of middle and small leaf tea group species or suitable improved varieties of tea tree collected from Qiannan Buyi and Miao Autonomous Prefecture of Guizhou Province and processed according to DB52/T 433. According to the raw material standard, combined with the processing technology and production practice, Duyun Maojian tea is divided into five grades: noble, precious, super, first and second.

尊品毛尖茶青（作者：司曼如）

经典 67

非遗传承·茶青采摘

生产一市斤尊品毛尖需要 4.2 市斤茶青，约 6 万至 6.5 万个芽头。尊品级都匀毛尖的茶青采摘时期，为每年的 3 月初至清明。采摘方法是提手采法，保证单芽，芽叶完整，叶色淡绿或深绿，叶质鲜嫩。

To produce a kilogram of top Maojian, it needs 4.2kg of tea leaves, about 120,000 to 130,000 bud heads. The picking period of the top grade Duyun Maojian tea is from the beginning of March to Qingming every year. The picking method is to pick by hands, so as to ensure that it is solitary buds, and buds and leaves are intact, light green or dark green, fresh and tender.

锅热达到400至450℃时，投放鲜叶。采用翻、抓、抛、抖、捻、搓等十多种手势，使茶青尽快达到80℃。眼看茶叶表面失去光泽，鼻闻茶香溢出，无生青味，手感茶粘手、柔软时，进行降火揉捻。揉捻至茶叶达到六成干时转为搓团。搓团至茶条卷曲，含水量达30%时进入提毫。提毫至茶条紧细卷曲、白毫显露时进行烘焙。烘焙至其含水量仅6%时起锅。挑选出细脆焦叶即得成品，正所谓：茶不离手，手不离锅，一锅到底，一气呵成，锅中取宝。

When the heat of the pot reaches 400 to 450 ℃, add fresh leaves. More than ten gestures such as turning, grasping, throwing, shaking, twisting and rubbing were used to make the tea leaves reach 80 ℃ as soon as possible. When you see the surface of the tea leaves lose luster, smell the tea fragrance overflow, no raw flavor, and feel the tea leaves sticky and soft, you can lower the fire and knead it. Roll until the tea leaves are 60% dry and turn into rolling mass. Roll mass until the tea strips are curled and the moisture content reaches 30%, and then enter the flowering pekoe procedure. When the tea strips are tight and curly and the pekoe are exposed, bake it. When baking to 6% moisture content, remove the tea from the pot. Pick out the crisp yellow leaves and you'll get the finished product.

绿茶仙子（作者：王殿维）

非遗传承·杀青

毛尖茶青（作者：王殿维 临）

都匀毛尖茶嫩芽杀青时，锅温在430℃左右，投叶量500 ~ 700克。以抖为主，抖焖结合，采用双手翻炒的手势。做到抖得散，翻得匀，杀得透。当叶质转软，清香透露，降低锅温进入揉捻工序。

When the tender buds of Duyun Maojian tea removing water, the pot temperature is about 430℃ , and the quantity of tea leaves is 500-700g. Mainly shake, shake and cover. Use the gesture of flipping with both hands. Shake it loose, turn over well, and remove water thoroughly. When the leaves turn soft, the fragrance is revealed, and the temperature of the pot is reduced to enter the rolling process.

毛尖揉捻（作者：王殿维）

揉时长，用力重，是都匀毛尖茶揉捻的特点，是形成毛尖茶味浓的因素之一。将锅温降至 65℃左右，用单手或双手沿锅边翻起茶叶置于双手中，视温度或揉转二周解块抖散一次，或揉转三、四周解块抖散一次，至基本成条卷曲，叶不黏手，易散开为度，达五六成干时即转入搓团提毫工序。

Long rolling time and heavy force, are the characteristics of the rolling of Duyun Maojian tea, which is one of the factors to form the strong taste of Maojian tea. Lowering the temperature of the pot to about 65℃ , turn up the tea leaves along the edge of the pot with one or both hands and put them in both hands. Depending on the temperature of the pot or knead and turn for two times ,or knead and turn for three or four times to remove the block and shake once. In this way, the tea leaves are basically into strips and curls, and the leaves do not stick to hands and are easy to disperse. When it reaches 50-60% dry, it will be transferred to the process of rolling mass and flowering pekoe.

非遗传承·搓团提毫

毛尖搓团提毫时锅温50 ~ 60℃，将茶叶握在掌中合掌旋搓，搓成茶团，抖散炒干，反复数次至七成干度。改用双手捧茶，压搓茶条，边搓边炒，搓炒结合，搓至白毫竖起。茶叶约八成至九成干时，降低锅温（50℃以下），将茶叶薄摊入锅中，炒至足干。炒干时做轻巧翻炒动作，使茶叶里外干度一致，增进香气。

毛尖搓团提毫（作者：王殿维）

The pot temperature is 50-60 ℃ when rolling mass and flowering pekoe. Hold the tea leaves in the palm and rub it together. Rub it into a tea mass. Jitter and spread, stir fry to dry. Repeat several times to 70% dryness. Instead, hold the tea leavs with both hands, press and rub the tea strips, stir fry while rubbing, combine the rubbing and frying until the pekoe stands up. When the tea leaves are about 80% to 90% dry, lower the pot temperature (below 50 ℃). Spread the tea leaves thin in the pot. Stir fry until dry. When stir dry, do light stir fry action, make tea inside and outside dry degree consistent, in order to enhance aroma.

清香满堂（作者：孔六庆）

经典 72

尊品干茶

都匀毛尖茶干茶外形白毫显露、条索紧细、卷曲似鱼钩。其农药残留和其他污染几乎为零，重金属含量也特别低。都匀毛尖茶的铅平均含量在 0.4 ~ 0.5mg/kg 之间，远远低于国家标准 2mg/kg。都匀毛尖茶内含的营养物质也较为丰富。贵州省茶科所测定，其茶多酚含量高达 31.24%，比一般茶叶约高 10%；氨基酸含量为 124.51 mg/g，咖啡碱含量为 28%，水浸出物含量为 38.21%，儿茶素含量为 124mg/g。

The appearance of Duyun Maojian dry tea is that the pekoe are exposed, the tea strips are tight and thin, curly like fishhooks. Its agricultural residue and other pollution are almost zero, and its heavy metal content is also very low. The average content of lead in Duyun Maojian tea is 0.4 - 0.5 mg/kg, which is far lower than the national standard of 2 mg/kg. Duyun Maojian tea is also rich in nutrients. According to the determination of the Tea Sience Research Institute of Guizhou Province, the content of tea polyphenols is up to 31.24%, which is about 10% higher than that of general tea; the content of amino acid is 124.51 mg/g; the content of caffeine is 28%; the content of water extract is 38.21%; the total content of catechin is 124 mg/g.

问君何时到山中 都匀毛尖茶已熟
德昌画

毛尖茶已熟（作者：范德昌）

经典 73

明前毛尖

都匀毛尖春茶有明前茶、谷雨茶和春尾茶三种。清明节（4月5日左右）前采制的茶是明前茶，这期间的茶叶嫩，喝着有种淡淡的香，几乎100%嫩芽头，是都匀毛尖中级别最高的茶。

The spring tea of Duyun Maojian has three kinds of Mingqian tea, Guyu tea and Chunwei tea. The tea produced before the Qingming Festival (around April 5th) is Mingqian tea. During this period, the tea is tender, with a light fragrance, almost 100% bud heads, which is the highest level of Duyun Maojian tea.

幸福平安（**作者：刘敏艳**）

经典 74
谷雨毛尖

谷雨（4 月 20 日左右）前采制的都匀毛尖茶。春季温度适中，雨量充沛，茶叶的生长正在含苞怒放，一芽一叶正式形成。泡好的“条形”虽然仅次于明前茶，但是味道稍微加重了。这种茶集都匀毛尖外形的档次感和它的口感为一体（外形、口感各占 50%），主要适合较高端消费人群。

Before Grain Rain (around April 20th), the Duyun Maojian tea is Maojian Guyu tea. The temperature is moderate and the rainfall is abundant in spring. The growth of tea is in full bloom, with one bud one leaf formally formed. Although the “strip” is second only to Mingqian tea, the flavor is slightly enhanced. This kind of tea integrates the sense of appearance and the sense of taste of Duyun Maojian tea (50% in appearance and 50% in taste respectively), which is mainly suitable for slightly high-end consumers.

经典 75 春尾毛尖

清品图（作者：王林）

春天末期前采制的都匀毛尖茶。经常说的雨前茶大概就是这个时期的茶。大家都知道，最好的茶要喝春天的！因为一个冬天的精华都体现在这个时候。口感也不例外！实际意义上的好茶也告一段落。这时候的茶叶价格已经成熟，条形虽然不能和明前、谷雨相比，但是绝对耐泡好喝，价位相对比较便宜，这种茶适合大众人群。

The Duyun Maojian produced before the end of spring is called Maojian Chunwei tea. It is often said that Yuqian tea is probably the tea of this period. We all know the best tea to drink in spring! Because the essence of a winter is reflected in this time. Taste is no exception! In a practical sense, good tea also comes to an end. At this time, the price of tea is mature. Although the strip shape can not be compared with the Mingqian and Guyu, it is absolutely resistant to brewing and good to drink, and the price is relatively cheap. This kind of tea is suitable for the public.

毛尖茶韵（作者：王长存 临）

经典 76

夏茶毛尖

夏天采制的都匀毛尖茶是夏茶毛尖。随着温度升高，雨水充足，茶叶也迅速生长，价格是一年最低的！叶子泡出来比较大、宽。茶水比较浓，味道微苦，耐泡，主要适合对茶叶需求比较大的饭店、茶馆等。因为价格相对便宜，喜欢喝浓茶的朋友，这个其实挺不错的。

The Duyun Maojian produced in summer is Maojian summer tea. With the increase of temperature and rainfall, tea grows rapidly. The price is the lowest in a year! The leaves are bigger and wider. The tea is relatively strong, slightly bitter in taste and resistant to brewing. It is mainly suitable for restaurants and teahouses with large tea demand. Because it's relatively cheap, for those who like to drink strong tea, this tea is actually quite good.

清品宜香（作者：王林）

经典 77

白露毛尖

古人说：“春茶苦，夏茶涩，要好喝，秋白露。”白露茶即秋茶，是二十四节气立秋至白露之间采摘的茶，又叫“谷花茶”。此时的茶叶经过夏季的酷热，白露前后正是它生长的极好时期。既不像春茶那样鲜嫩，不经泡，也不像夏茶那样干涩味苦，而是有一种独特的甘醇清香味，尤受茶客喜爱。

The ancients said: “spring tea is bitter, summer tea is astringent. It’s good to drink, autumn Bailu.” Bailu tea is autumn tea,which is picked from the begining of autumn to Bailu with 24 solar terms. It is also called “Guhua tea.” At this time, the tea has gone through the summer heat. Before and after Bailu, it is a good time for its growth. This tea is neither as fresh and tender as spring tea, nor as dry and bitter as summer tea, but has a unique sweet and mellow fragrance, which is particularly popular with tea customers.

毛尖沏茶

毛尖茶香（作者：王殿维 临）

都匀毛尖茶采用“高水温，多投茶，快出汤，茶水分离，不洗茶”的“贵州冲泡法”。这种便捷、简单、高效而又不乏仪式感的冲泡方式，很快普及到寻常百姓家，正在全国风靡流行。用100℃开水冲泡的毛尖，香气高，茶条舒展迅速，茶叶中的内含物质溶解最快，最能集中体现都匀毛尖“香高味浓”的特点。

The “Guizhou brewing method” of “high water temperature, more tea, quick soup, separation of tea and water, no tea washing” is adopted for brewing Maojian tea. This convenient, simple, efficient and ritualistic way of brewing quickly spread to ordinary people’s homes and is becoming popular all over the country. The Maojian tea brewed with 100 degree boiling water has high aroma, rapid stretching of the tea strips, and the fastest dissolution of the contents in the tea, which can best reflect the characteristics of Duyun Maljian with “high fragrance and strong taste”.

经典79

毛尖赏茶

毛尖赏茶（作者：王殿维）

毛尖沏泡之前，先请客人欣赏待泡茶叶的形状、颜色，干闻香气。主客边看边交谈赏茶印象，而后用玻璃杯或白瓷盖碗冲泡。用玻璃杯可观察到茶在水中上下翻腾，缓缓舒展，穿梭游动，犹如茶仙子曼妙飞舞，美轮美奂。

Before brewing, please enjoy the shape and color of the tea to be soaked and smell the fragrance. The host and guest talk and enjoy the tea impression while watching, and then make it with glass or white porcelain covered bowl. With a glass, you can see the tea rolling up and down in the water, slowly stretchjing, shuttling and swimming, just like the tea fairy dancing, beautiful.

毛尖茶汤（作者：司曼如）

经典80

毛尖茶汤

在中国的绿茶领域，贵州绿茶是上上品，都匀毛尖则是贵州绿茶中的极品、奢侈品。都匀毛尖茶汤色清澈明亮，素以“三绿三黄”的品质特征著称于世：干茶绿中带黄，汤色绿中透黄，叶底绿中显黄。茶汤内质香高持久，滋味鲜爽回甘。

In the field of green tea in China, Guizhou green tea is the top grade, while Duyun Maojian is the best and luxury in Guizhou green tea. The color of Duyun Maojian tea soup is clear and bright. It is famous for its quality features of "three green three yellow": dry tea is green with yellow, soup color is green penetrating yellow, and the bottom of leaves are green showing yellow. The tea soup is fragrant and long-lasting, fresh and sweet.

经典81 毛尖闻香

毛尖闻香（作者：王殿维）

对于毛尖茶香的品鉴一般要三闻。一是闻干茶的香气（干闻），二是闻泡开后充分显示出来的茶的本香（热闻），三是闻茶香的持久性（冷闻）。闻香品茶，本就如此自然。时间如水般从容而过，内心遂平静、清凉，一片安宁。

Generally, there are three steps to smell fragrance of Maojian. The first is to smell the aroma of dry tea (dry smell), the second is to smell the original fragrance of the tea which is fully displayed after it is brewed (hot smell), and the third is to smell the persistence of the tea fragrance(cold smell). It's so natural to smell and taste tea. Time passes calmly like water, and the heart is calm, cool and peaceful.

经典 82

毛尖敬茶

毛尖敬茶（作者：舒肖）

毛尖敬茶是礼仪中待客的一种日常礼节。上茶时应以右手端茶，从客人的右方奉上。敬客斟茶通常以斟半杯为礼貌，俗称 " 茶七酒八 "，注意茶水不宜太烫。有两位以上的访客时，用茶盘端出的茶汤，每杯茶色要一致，并要左手捧着茶盘底部，右手扶着茶盘的边缘。如有茶点，应放在客人的右前方，茶杯应摆在点心右边。壶中茶叶可反复浸泡 3 至 4 次，客人杯中茶饮尽，主人可为其续茶，客人散去，方可收茶。

Serving Maojian tea for guest is a daily etiquette. Tea should be served with the right hand and served from the right side of the guest. It's usually polite to serve half a cup of tea to quests. It's commonly known as "tea seven wine erght". Pay attention that the tea should not be too hot. When there are more than two visitors, the color of each cup of tea soup brought out by the tea tray shall be the same, and the bottom of the tea tray shall be held in the left hand, and the edge of the tea tray shall be held in the right hand. If there is any refreshment, it should be placed in front of the right side of the guest, and the cup should be placed on the right side of the refreshment. The tea in the pot can be repeatedly soaked 3 to 4 times. When the tea in the guest's cup is exhausted, the host can refill for the guest. Only when the guest leaves, can the tea be collected.

溪山会友（作者：陈常淑）

经典 83

螺蛳壳毛尖产区

出产优质毛尖茶的螺蛳壳茶园，海拔 1580 米，中央为高寨水库，四周群山环抱，绿树成荫，系高档毛尖茶主要产区之一。茶区负离子浓度年平均值为每立方厘米 3908.2 个，联合国卫生组织认定空气清新的标准是空气中负离子含量达到每立方厘米 1500 个。

The Luosike tea garden, which produces high-quality Maojian tea, is 1580 meters above sea level, with Gaozhai reservoir in the center, surrounded by mountains and trees. It is one of the main production areas of high-grade Maojian tea.The annual average value of anion concentration in the tea area is 3908.2 per cubic centimeter. The standard of air freshness determined by the United Nations Health Organization is that the content of anion in air reaches 1,500 per cubic centimeter.

茶艺图（作者：李国锰）

经典 84

都匀毛尖（国际）茶人会

都匀毛尖（国际）茶人会是都匀毛尖引领“黔茶出山”的平台，是茶产业的盛会，也是展示茶产业的窗口，更是促进茶产业发展的机会。该会每年举行一届，以茶兴业、以茶惠民、以茶养文。茶人会永恒的宗旨是，热忱欢迎远方茶客，论道会茶都，观生态之州，游幸福黔南。

Duyu Maojian (international)Tea People's Conference is a platform for Duyun Maojian to lead "Qian tea out of the mountain", a grand conference of the tea industry, a window to show the tea industry, and an opportunity to promote the development of the tea industry. The conference is held once a year to develop business, benefit people and cultivate culture with tea. The eternal tenet of the tea people's conference is to warmly welcome tea guests from afar, discuss about Tao in tea capital, view the ecological prefecture, and travel to the happy Qiannan.

经典 85

“都匀毛尖杯”斗茶大赛

2014年以来，黔南州已连续举办了五届都匀毛尖春季斗茶大赛、一届秋季斗茶大赛。参赛企业和茶样一年比一年多，品质一届比一届好，影响力也是越来越大。这样的斗茶大赛得到了业界的普遍认同和欢迎，已经成为黔南茶人、茶企相互学习、相互切磋技艺的一个专业性竞技平台。

Since 2014, Qiannan Prefecture has held five consecutive Duyun Maojian tea competition in spring and one in autumn. The competition enterprises and tea samples are more and more year by year, the quality is better and the influence is more and more. Such a tea contest has been generally recognized and welcomed by the industry, and has become a professional competition platform for tea people and tea enterprises in Qiannan to learn from each other.

斗茶（作者：王殿维 临）

经典 86

『味精茶』都匀毛尖

客饮毛尖茶香（作者：周艺）

精明的浙江、江苏、福建、安徽茶商，年年通过“绿茶天路”，把收购到的都匀毛尖茶茶青或茶叶用飞机运回，共享当地茗茶盛誉。其中缘由，中国著名茶叶专家高端荣一语道破：“都匀毛尖茶长期以来作为‘味精茶’与高端茶匹配出口，深受欧美、日本客户的欢迎。”

The astute tea merchants in Zhejiang, Jiangsu, Fujian and Anhui, through the “sky road of green tea” every year, transport the purchased Duyun Maojian tea or tea leaves back by plane, and mix it with the local tea to sell or sell it directly under the brand, so as to earn a good price difference. The reason is that Gao Duanrong, a famous tea expert in China, lays bare its secret with one remark: “For a long time, Duyun Maojian tea has been exported as ‘monosodium glutamate tea’ matched with high-end tea, which is very popular with customers in Europe, America and Japan.”

清品图（作者：王林）

经典 87

都匀毛尖的六个全国之“最”

在中国的绿茶产地中，都匀毛尖茶产区海拔最高，降水最均匀，云雾最多，气候最温和，森林覆盖率最高，风景名胜数量最多。科学研究表明，海拔越高，茶树新梢中茶多酚和儿茶素含量越低，而氨基酸和芳香物含量越高。得天独厚的“天无三日晴”使都匀毛尖茶树新梢叶片大，持嫩性强，内含物丰富，节间长，产量高，品质好。云雾多、漫照光丰富，能促进茶叶内蛋白质和含氮芳香物的形成和积累。

Among the green tea producing areas in China, Duyun Maojian tea producing area has the highest altitude, the most uniform precipitation, the most clouds and mists, the mildest climate, the highest forest coverage and the largest number of scenic spots. Scientific research shows that the higher the altitude, the lower the content of tea polyphenols and catechins in tea shoots, and the higher the content of amino acids and aromatics. The unique "no three sunny days" makes the leaves of the new shoots of Duyun Maojian tea tree larger, with strong tenderness, rich contents, long internode, high yield and good quality. There are many clouds and abundant diffuse light, which can promote the formation and accumulation of protein and nitrogen-containing aroma in tea.

毛尖茶诗

"Good tea is always like a beautiful woman ". The way of poetry and tea is perfect. China is the country of poetry and the hometown of tea. The meeting of the two makes the most wonderful tea poetry witticism, both refreshing and reassuring. Over 20000 tea poems, tea CI and tea songs have been written for tea by Chinese literati and poets of all ages, but they are endless in chanting, composing and singing.

"从来佳茗似佳人"。诗茶之道，至善至美。中国是诗的国度，也是茶的故乡。两者相遇，造就了最美的茶诗妙语，既爽口，又爽心。我国历代文人墨客为茶写下了两万多首茶诗、茶词、茶曲，乃咏之不尽，赋之不绝，唱之不断啊。

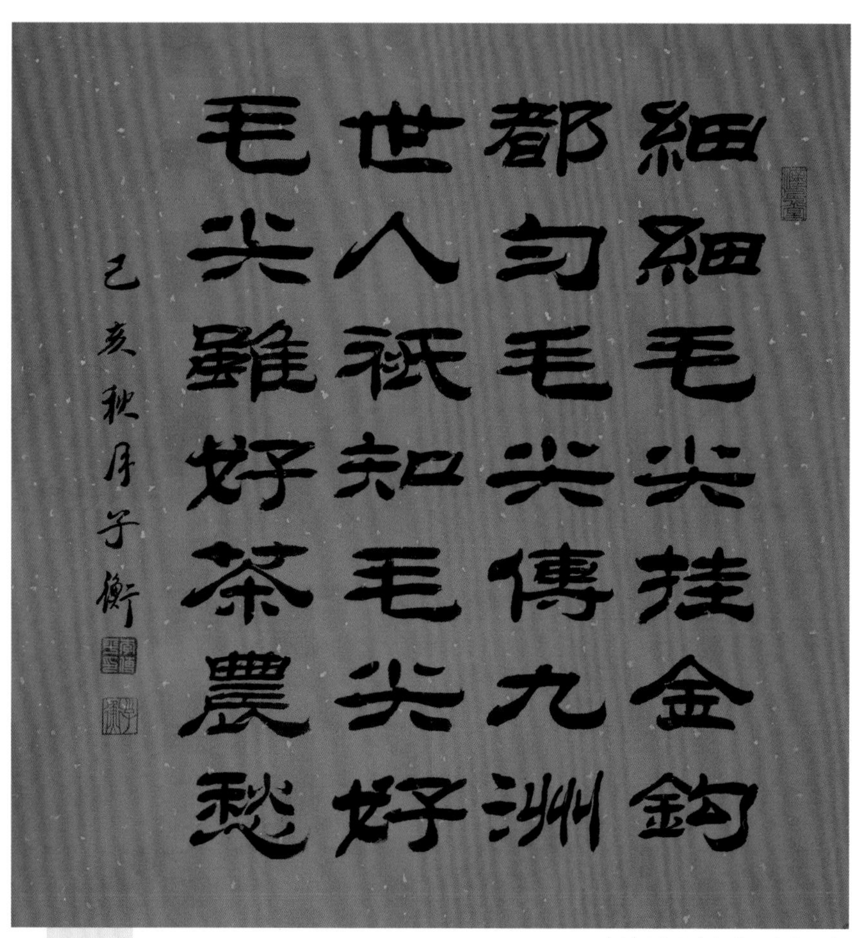

布依民谣（**书法：李传平**）

经典88

布依族民谣

“细细毛尖挂金钩，都匀毛尖传九洲，世人只知毛尖好，毛尖虽好茶农愁。”这是布依族世代相传的一首民谣，道出了都匀毛尖茶形如金钩，清香淡雅，名声远扬，被人称赞的溢美之词。同时也透露了从前茶农内心的忧愁。

This is a folk song handed down from generation to generation by the Buyi people. It expresses the praises of Duyun Maojian tea, which is like a golden hook, fragrant and elegant, well known and praised. At the same time, it also revealed the inner sorrow of the former tea farmers.

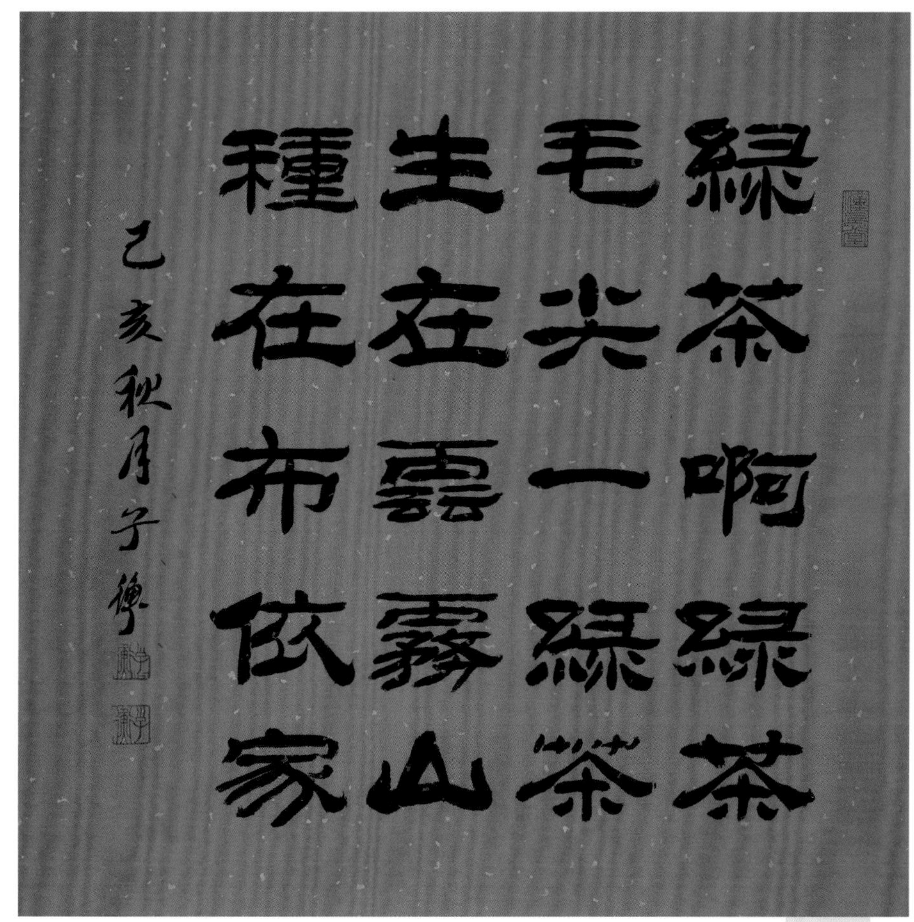

綠茶啊綠茶
毛尖一綠茶
生在雲霧山
種在布依家
己亥秋月子傳

仙女采茶舞诗歌（**书法：李传平**）

经典 89 《仙女采茶舞》

这是都匀毛尖茶传说里的一首诗。蛮王的九十个姑娘在得到绿仙雀变成的漂亮茶姐姐面授机宜时，一阵欢笑，高兴得边跳边唱《仙女采茶舞》。诗词内容是：“绿茶啊！绿茶，毛尖一绿茶。生在云雾山，种在布依家。”

This is a poem in the legend of Duyun Maojian tea. When the 90 girls of king were taught skills by the beautiful tea elder sister that the green fairy bird became, they burst into laughter and danced and sang “Fairy Picking Tea Dance”.

经典 90

都匀《敬茶歌》

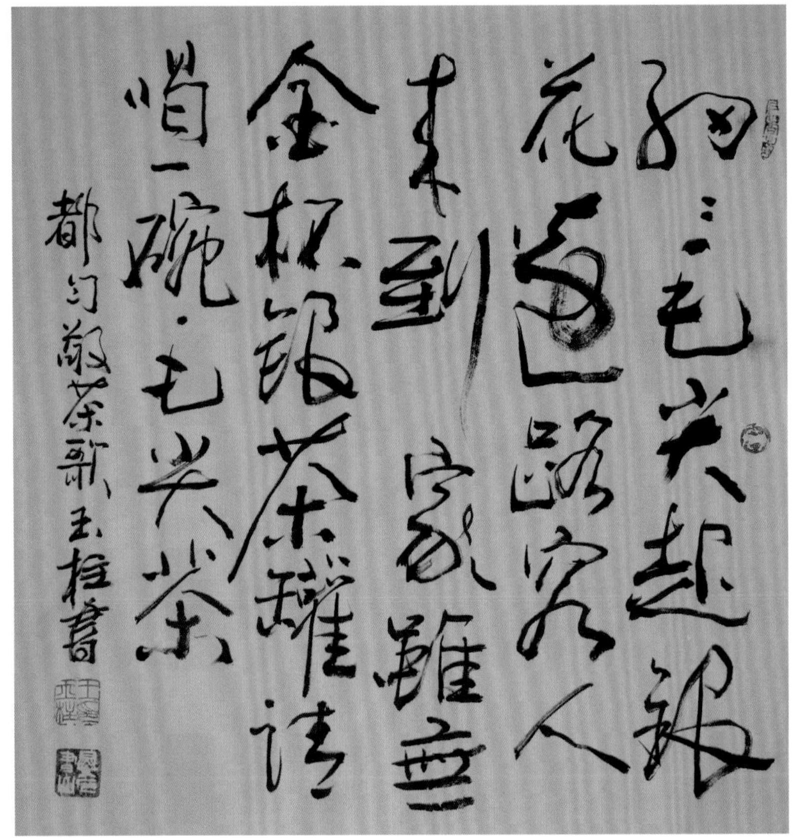

都匀《敬茶歌》（书法：王玉柱）

“细细毛尖起银花，远路客人来到家。虽无金杯银茶罐，请喝一碗毛尖茶。”这首诗是都匀人民热情友好，以茶待客，以茶会友，以茶联谊，以茶示礼，最真挚最生动、最直白的具体表现。

This poem is the most sincere, vivid and straightfoward expression of Duyun people’s enthusiasm and friendliness, treating guests with tea, meeting friends with tea, keeping up a friendship with tea, showing ceremony with tea.

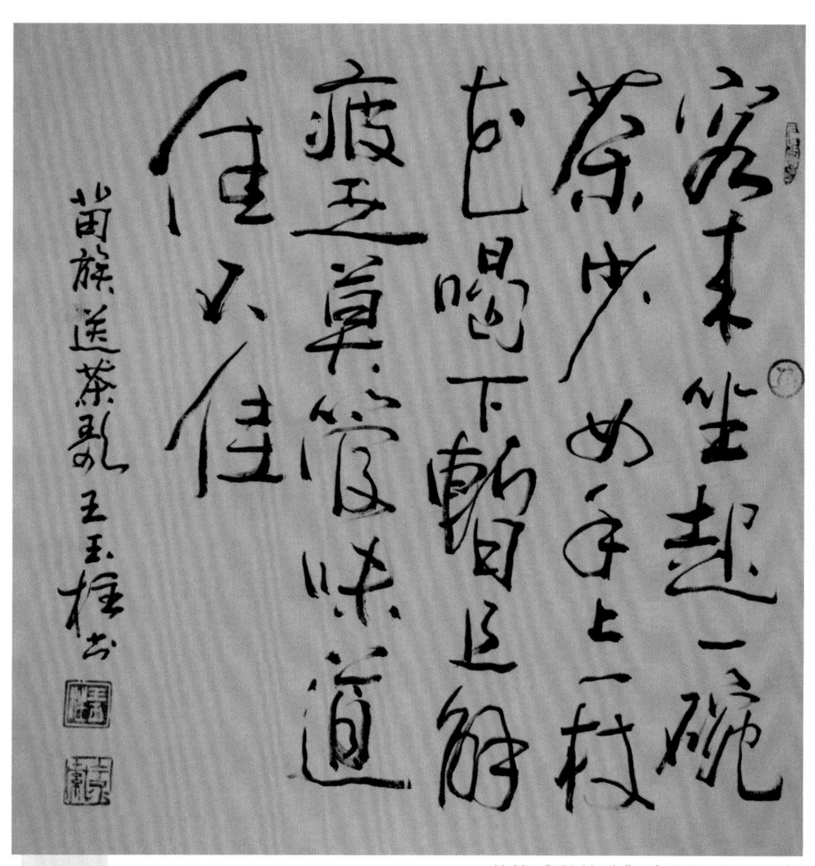

苗族《送茶歌》（书法：王玉柱）

经典 91

苗族《送茶歌》

“客来坐起一碗茶，少女手上一枝花，喝下暂且解疲乏，莫管味道佳不佳。”

黔南的苗族同胞有着悠久的种茶、喝茶史，喝茶成俗，并将茶作为寄托或表达思想情感乃至道德观念的载体代代相传。在他们的衣食住行、婚丧嫁娶、生老病死、节庆文娱等社会交往中，处处离不开茶。

Miao compatriots in Qiannan have a long history of tea planting and drinking, and tea has been passed down from generation to generation as a carrier of sustenance or expression of thoughts, emotions and even mortal concepts. In their social contacts, such as basic necessities of life, weddings and funerals, life and death, festivals and entertainment, tea is indispensable everywhere.

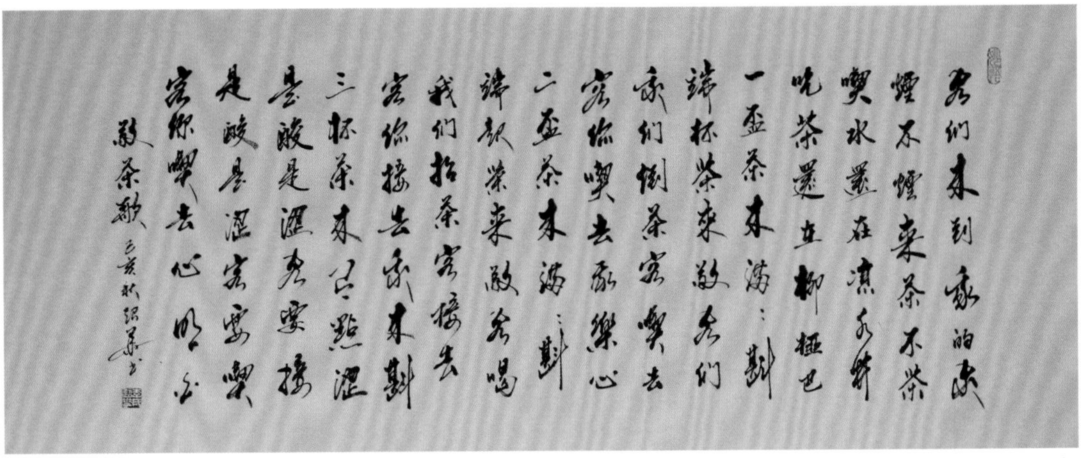

贵定《敬茶歌》（书法：李超华）

经典 92

贵定《敬茶歌》

客们来到我的家，烟不烟来茶不茶，

吃水还在凉水井，吃茶还在树桠巴。

一杯茶来满满斟，端杯茶来敬客们，

我们倒茶客吃去，客你吃去我乐心。

二杯茶来满满斟，端起茶来敬客喝，

我们抬茶客接去，客你接去我斟来。

三杯茶来有点涩，是酸是涩客要接，

是酸是涩客要吃，客你吃去心明白。

贵定县有 2000 多年的种茶史，600 多年的贡茶史。贵定云雾茶（属都匀毛尖品牌）在唐、宋、元、明、清时均为朝廷贡茶，属清朝八大名茶，是贵州唯一、全国罕见、既有史志记载又有碑文铭刻的贡茶。

There are more than 2000 years of tea planting history and more than 600 years of tribute tea history in Guiding County of Qiannan. In Tang, Song, Yuan, Ming and Qing Dynastys, Yunwu Tea in Gguiding was the tribute tea of the imperial court, which was one of the eight famous tea of Qing Dynasty. It was the only tribute tea in Guizhou, which was rare in China and has both historical records and inscriptions.

经典 93

庄晚芳教授毛尖茶诗

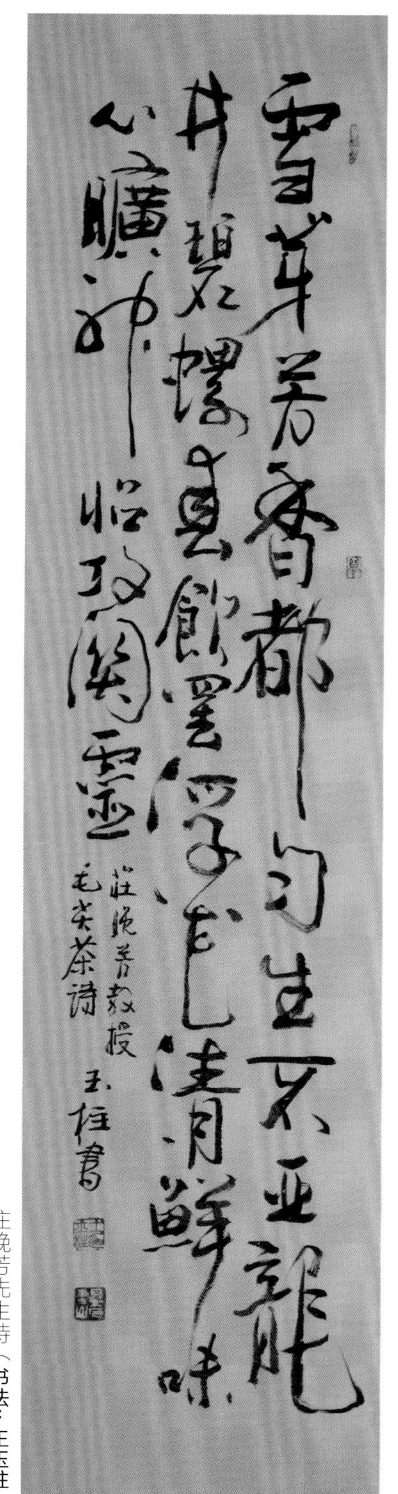

庄晚芳先生诗（书法：王玉柱）

“雪芽芳香都匀生，不亚龙井碧螺春。饮罢浮花清鲜味，心旷神怡攻关灵。”

庄晚芳教授是我国茶树栽培学科的奠基人之一。20 世纪 60 年代末，都匀对毛尖茶的旧有工艺进行改革。经新工艺制作的都匀毛尖茶加工出来后，都匀茶场给他寄去了一包样品。庄晚芳先生品尝后，评价很高，于是赋诗夸赞。

Professor Zhuang Wuanfang is one of the founders of tea cultivation discipline in China. In the late 1960s, Duyun reformed the old techology of Maojian tea. Duyun Maojian tea made by the new technology was processed, and Duyun tea farm sent him a package of samples. After Mr. Zhuang Wanfang tasted it, he spoke highly of it and praised it with poems.

经典 94

张大为赋诗赞毛尖

“不是碧螺，胜似碧螺。香高味醇，别具一格。”

张大为先生是中国茶叶文化宣传的“先驱者”。他极力倡导“俭、美、和、静”的中国茶德，并促成了中国首套（四枚）茶文化邮票的发行。在品评都匀毛尖后，他欣然题诗赞美。

Mr. Zhang Dawei is the “pioneer” of Chinese tea culture publicity. He strongly advocated the “thrifty, beautiful, harmonious and quiet” Chinese tea morality, and contributed to the issuance of China’s first set of (four) tea culture stamps. After tasting and appraising Duyun Maojian tea, he happily wrote poems and praises.

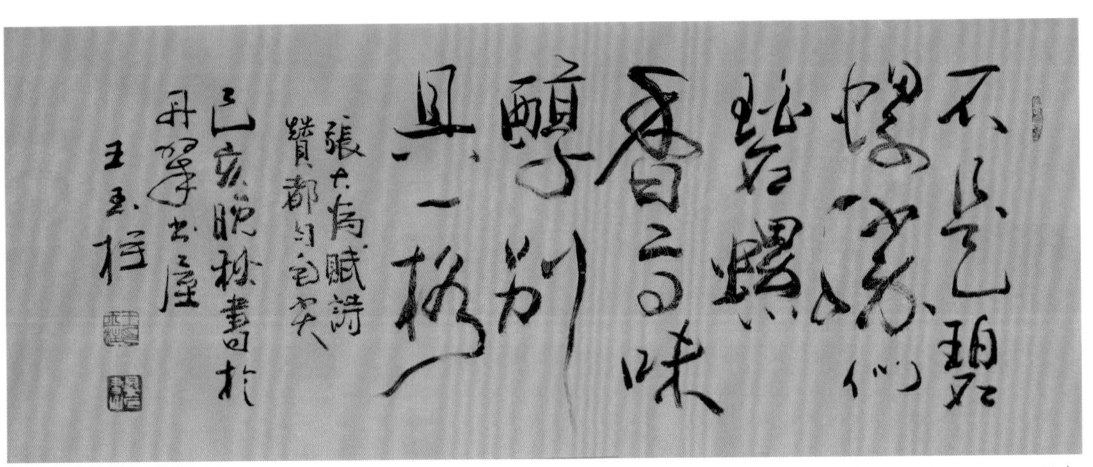

张大为先生诗（书法：王玉柱）

经典 95 北京诗人咏都匀毛尖

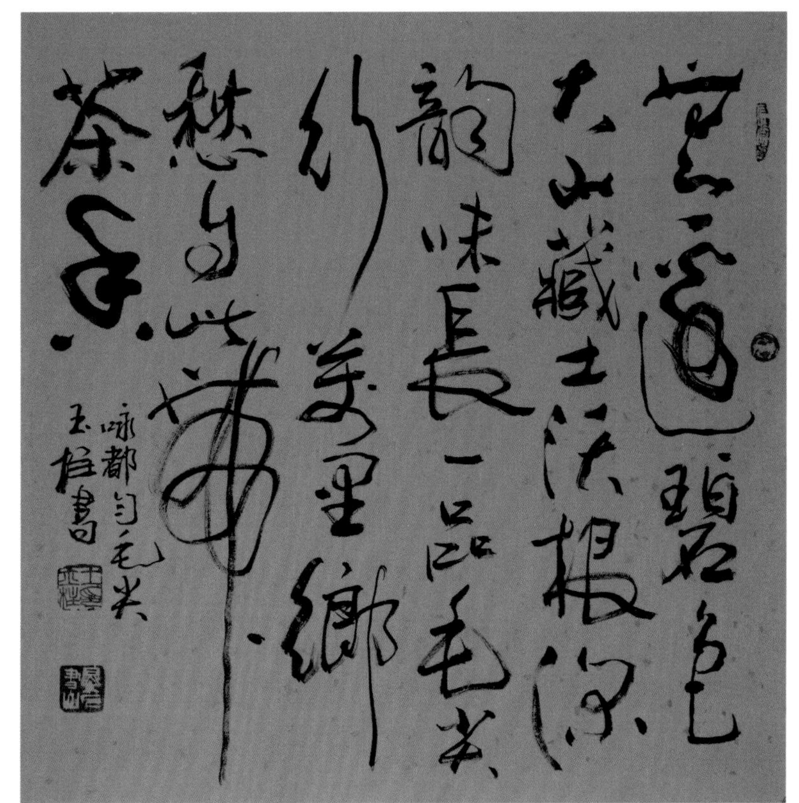

李树喜先生诗（书法：王玉柱）

无边碧色大山藏，土沃根深韵味长。

一品毛尖行万里，乡愁自此带茶香。

2017 都匀毛尖（国际）茶人会首届“都匀毛尖杯”中华茶诗词大赛在毛尖茶都都匀市举办。仅一个月的征稿时间，收到参赛作品 3856 首。经评审组层层筛选与评定，评出特等奖 1 名、一等奖 1 名、二等奖 2 名、三等奖 3 名、优秀奖 60 名、入围奖 273 名。这首诗的作者是北京的李树喜，三等奖获得者。

The first "Duyun Maojian Cup" Chinese Tea Poetry Competition of 2017 Duyun Maojian (international) Tea People's Conference was held in Duyun City, Maojian tea capital. Only one month's draft time, 3856 entries were received. After selection and evaluation by the evaluation team, one special prize, one first prize, two second prizes, three third prizes, 60 excellent prizes and 273 shortlisted prizes were awarded. The author of this poem is Li Shuxi from Beijing, the third prize winner.

经典96

水书写意都匀毛尖

云上心茶（书法：蒙富春）

在中国的少数民族文字中，黔南州水族人民的水书是一种很奇特的古老文字，记载了水族的天文、地理、民俗、伦理、哲学、美学等文化信息，被称为水族的“百科全书”。水书共有400多个单字，类似甲骨文和金文，被称为象形文字的“活化石”，2006年被列为国家级非物质文化遗产。这幅用水书表现的都匀毛尖茶语，由一生心系毛尖的水书习俗传承人蒙富春书写，意为：云上心茶。

Among the ethnic minority languages in China, Shuishu in Sandu County, Qiannan Prefecture is a very peculiar ancient writing, which records the astronomical, geographical, folk custom, ethics, philosophy, aesthetics and other cultural information of the Shui nationality, which is called the "Encyclopedia" of Shui nationality. There are more than 400 words in Shuishu, which are similar to Oracle Bone inscriptions and gold inscriptions. It is called "living fossil" of hieroglyphics. In 2006, it was listed as a national intangible cultural heritage. The tea language of Duyun Maojian expressed in the Shuishu was written by Meng Fuchun, an inheritor of Shuishu custom who had been concerned about Maojian all his life. The meaning of it is: Favorit tea on the cloud.

毛尖茶人

Duyun maojian from "treasure in the fire", from picking, removing water, rolling, rolling mass and lifting pekoe, even pacaging, transportation, publicity, drinking, every step is permeated with Maojian tea people's understanding of tea, giving the flavor of tea emotion, carrying the deep feelings of the Chinese nation.

来自“火中取宝”的都匀毛尖茶，从采摘、杀青、揉捻、搓团提毫，甚至包装、运输、宣传、品饮，每一步都浸润着毛尖茶人对茶的理解，赋予茶情感的韵味，承载着中华民族的深深情结。

毛尖茶专著《鱼钩巷》主编魏明禄

魏明禄（作者：郑锋）

黔南州政协党组书记、主席魏明禄，获工学博士学位和项目管理副教授职称，学术成果丰厚。其茶学专著除《鱼钩巷》外，另有《黔南茶树种质资源》一书。以往曾在《人民论坛》《华中科技大学学报》等刊物发表论文近 30 篇，出版了专著《公共危机应对机制研究》。主编的《王学胜地》一书获贵阳市哲学社会科学成果三等奖。

Wei Minglu, secretary and chairman of the Political Consultative Conference of Qiannan Prefecture, has a doctor's degree in engineering and an associate professor of project management. In addition to Fish Hook Lane, his monograph on tea science is another book of Germplasm Resources of Qiannan Tea.

钟爱都匀毛尖茶的『国家一级美术师』黄天虎

黄天虎（左）与方兴齐（作者：郑锋）

中国美术家协会会员、贵州省美术家协会前常务理事、贵州省中国画艺委会主任黄天虎，艺术造诣精湛，成果丰硕，作品多次参加全国美术作品展，十余件作品在全国和省内获奖，不少作品在美国、日本、新加坡等国展出，很多编入各地大型画册，数百件见诸报刊。黄天虎热爱都匀毛尖茶，曾深入都匀毛尖茶产地黄河、大槽等考察调研，不遗余力宣传都匀毛尖茶。

Huang Tianhu, member of China Artists Association, former executive director of Guizhou Artists Association and director of Guizhou Chinese Painting Art Committee, has exquisite artistic attainments and fruitful achievements. His works have participated in national art exhibitioans for many times. More than ten works have won awards in China and the province. Many of them have been exhibited in the United States, Japen, Singapore and other countries. Many of them have been included in large scale picture albums around the country, and hundreds of them have been published in newspapers and magazines. Huang Tianhu loves Duyun Maojian tea. He once went to Huanghe, Dacao and other places whtere Duyun Maojian tea was produced. He carried out investigation and research and spared no effort to promote Duyun Maojian tea.

唐元稹诗（书法：苏意明）

经典 99

《中国茶道诗词书法集》作者苏意明

都匀市书法家协会艺术顾问、黔南州诗词楹联协会顾问苏意明，一生爱茶，以自己五十余载文学艺术造诣，耗时十年有余，阅遍《全唐诗》《全宋词》《全元曲》《清诗别裁》《黔诗记略》等无数古今名卷，从数万作品中遴选出涉茶优秀诗词句对联 4120 首，撰成 74 册、3564 页鸿篇巨著：《中国茶道诗词书法集》。功在当代，利在千秋。

Su Yiming, art consultant of Duyun Calligrapher Association and consultant of Qiannan Poetry Couplet Association, love tea all his life. With more than 50 years of literary and artistic attainments, he spent more than 10 years reading numerous ancent and modern famous volumes such as Quan Tang Poetry, Quan Song Ci, Quan Yuan Qu, Qingshi Biecai, Qianshi Jilue, and so on. He selected 4120 excellent poems, ci, sentences and couplets related to tea from tens of thousands of works, and wrote 74 volumes and 3564 pages of masterpieces: Colligraphy Collection of Chinese Tea Ceremony Poetry. The achievements are in the present age, and the benefits are in the future.

经典 100

都匀毛尖现代工艺创始人徐全福

徐全福（作者：郑锋）

都匀茶场原场长、高级农艺师徐全福，是都匀毛尖现代加工工艺及都匀毛尖品牌创始人。1982 年，徐全福带着亲手加工的都匀毛尖茶赴长沙，参加商务部主办的全国首届名茶评比活动，从 86 个茶样中脱颖而出位列前十，被评为中国十大名茶，从此名震全国。2017 年荣获全国第三届“觉农勋章”奖。

Xu Quanfu, former head of Duyun tea farm and senior agronomist, is the founder of modern processing technology and brand of Duyun Maojian. In 1982, Xu Quanfu went to Changsha with the Duyun Maojian tea processed by himself to participate in the first national famous tea competition sponsored by the Ministry of Commerce. It stands out from 350 tea samples and ranks in the top ten, and is rated as the top ten famous tea in China. Since then, Duyun Maojian tea is famous all over the country. In 2017, he won the third national “Juenong Medal” award.

经典 101

都匀毛尖“把关人”欧平勇

欧平勇（作者：郑锋）

黔南州农科院副院长欧平勇，和茶叶打了 20 多年交道，参与了都匀毛尖茶从种植到加工制作的全过程，是名副其实的都匀毛尖“把关人”，曾获得“贵州省十大制茶能手”等 20 多项荣誉称号。

Ou Pingyong, vice president of Qiannan Academy of Agricultural Sciences, has been dealing with tea more than 20 years and participated in the whole process of Duyun Maojian tea from planting to production. He is a real “gatekeeper” of Duyun Maojian tea, and has won more than 20 honorary titles such as “top ten tea makers in Guizhou Province”.

经典 102

都匀毛尖制作技艺非遗传承人张子全

张子全（作者：郑锋）

中国文化部第五批国家级非物质文化遗产传承人张子全，世代以制茶贩茶为生。12岁炒制人生第一锅毛尖茶，飘香四溢，深受好评。一生与茶打交道，对工序、火候拿捏精准，工艺独到，并深知茶产业的发展离不开茶文化的发掘与弘扬。

Zhang Ziquan, the fifth group of national intangible cultural heritage inheritors of the Ministry of Culture, has been making and selling tea for generations. At the age of 12, he made the first pot of Maojian tea, which was fragrant and well received. He has been dealing with tea all his life. He has a precise understanding of the process and the fire. He has unique technology. He knows very well that the development of tea industry cannot be separated from the excavation and promotion of tea culture.

经典 103

都匀非公茶企联合党支部书记韦洪平

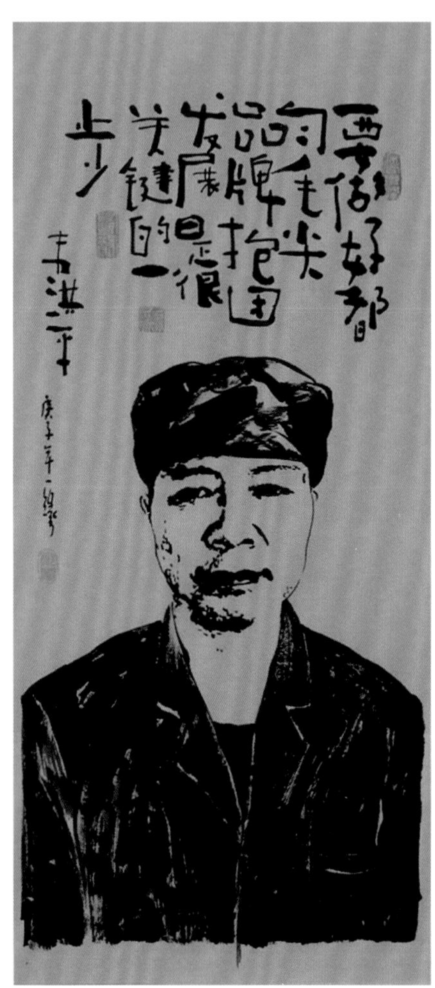

韦洪平（作者：郑锋）

都匀匀城春茶叶有限公司总经理韦洪平做茶近 30 年，创建了都匀毛尖“匀城春”品牌，产品质量达到欧盟标准，曾获“贵州省名优绿茶”称号、“恒天杯”全国名优绿茶评比“金奖”“中国名茶”评选金奖、“贵州省名牌产品”称号。他不忘初心，激励和带领周边党员群众发展茶产业，直接带动 211 户茶农参与茶园管理，让 682 户 2561 名村民通过茶叶经营真正吃上了“茶叶饭”，实现脱贫致富。

Wei Hongping, general manager of Duyun Yuncheng Spring Tea Co.,Ltd., has been making tea for nearly 30 years, and has established Duyun Maojian "Yuncheng Spring" brand. The product quality has reached the European Union standard, and has won the title of "Guizhou famous green tea", "Hengtian Cup" national famous green tea evaluation "Gold Award", "China famous tea" selection Gold Award, the title of "Guizhou famous brand product" and so on. He did not forget his original intention, encouraged and led the surrounding party members and the masses to develop the tea industry, directly led 211 tea farmers to participate in the management of the tea garden, and enabled 682 families and 2561 people to truly eat the "tea rice" through the tea business and get rid of poverty and become rich.

经典 104

都匀毛尖『十佳匠心茶人』蔡邦红

蔡邦红（作者：郑锋）

贵州都匀毛尖茶叶有限公司总经理蔡邦红，2000 年荣获都匀市首届十佳青年称号。30 多年来，一直致力于都匀本地原生茶树种的培育与加工、附产品开发利用。先后取得四个实用新型专利技术，初步形成了“产、学、研、农、工、贸”产业一体化发展模式，成为贵州省最早一家向欧盟、美、日等出口茶产品的企业。

Cai Banghong, general manager of Guizhou Duyun Maojian Tea Co., Ltd., won the title of the first ten best youth in Duyun City In 2000. For more than 30 years, He has been committed to the cultivation and processing of native tea species in Duyun, as well as the development and utilization of by-products. He has successively obtained four utility model patent technologies, initially formed the industrial integration development mode of "production, learning, research, agriculture, industry and trade", and became the first enterprise in Guizhou Province to exprot tea products to the European Union, the United States and Japan.

陈元安（作者：郑锋）

经典 105

从医者到茶人的陈元安

从事茶叶加工、销售及进出口贸易近 30 年的黔南苗岭工贸有限公司董事长陈元安，毛尖茶产业做得风生水起，获奖无数，只缘于他始终把茶界老前辈“做名优茶一定要有好的品种、好的种植环境和好的加工工艺”的教导，作为自己做好都匀毛尖茶一生追求的坚定信念。

Chen Yuan’an, chairman of Qiannan Miaoling industry and Trade Co. Lid., who has been engaged in tea processing, sales and import and export trade for nearly 30 years, has won numerous awards for his excellent Maojian tea business, only because he always takes the teaching of “to be famous and excellent tea, we must have good varieties, good planting environment and good processing technology” as his firm belief to be good at Duyun Maojian tea.

品茗图（作者：黄天虎）

经典 106

“黄河人”方兴齐

从小受父辈族人种茶、制茶、品茶熏陶的方兴齐，2006年创建了都匀毛尖茶品牌“黄河人”。长期以来，他始终秉承着茶叶品质三大要素“产地、品种、工艺”的制茶理念，继承了古法种、新法制中“茶叶须大小均匀，受热均匀，受力均匀”的核心技术，并在研制出“黄河人”毛尖茶专用电炒锅后，终于让自己的茶叶产品品质更加稳定，达到了“外形细紧卷曲、汤色黄绿明亮、栗香高锐持久、滋味鲜爽回甘、叶底嫩匀黄亮”的品质要求，深受广大茶人的喜爱和推崇。

Fang Xingqi, who was influenced by his father's generation of tea planting, tea making and tea tasting, founded the Duyun Maojian tea brand "Yellow River People" in 2006. For a long time, he has always adhered to the tea making concept of "origin, variety and technology", which are the three major elements of tea quality, and inherited the core technology of "uniform size, uniform heating and uniform stress". After the development of the "Yellow River People" Maojian tea special electric frying pan, the quality of his tea products is more stable, reaching the quality requirements of "thin and tight curly shape, bright yellow and green soup color, high and durable chestnut fragrance, fresh and refreshing taste, tender and even yellow leaf buttom", which is deeply loved and respected by the majority of tea people.

经典 107

“匀东布依妹”王国叶

王国叶（作者：郑锋）

身为都匀市毛尖企业商会秘书长、广东援黔企业家联合会会员、土生土长返乡创业的布依族美女企业家王国叶，带领 30 多个异姓结拜姊妹及都匀市匀东镇新坪村 200 多户布依乡亲种茶致富，打造出 " 匀东布依妹 " 都匀毛尖这块特色品牌，让家乡 4000 多亩近乎荒废的茶山重焕生机，为村民带来致富希望。

Wang Guoye, the Secretary-General of Maojian Enterprise Chamber of Commerce in Duyun City, a member of Guangdong Federation of Entrepreneurs in aid of Guizhou Procince, and a native businesswoman of Buyi Nationality who returned home to start her own business, led more than 30 sworn sisters of different surnames and more than 200 families of Buyi Nationality in Xinping Village, Yundong Town, Duyun City to grow tea and become rich. She has created the characteristic brand of “Yundong Buyi Sister” Duyun Maojian, to revive the nearly deserted tea mountain of more than 4000 mu in her hometown and bring hope for the villagers to become rich.

旅游商品传播者李尚飞

李尚飞（作者：郑锋）

李尚飞现为贵州省都匀市名品斋旅游商品开发有限责任公司法人代表，从事经营都匀毛尖茶、民族手工艺品、旅游商品 20 余载，热心保护和开发本地野生茶树种事业，产品销往国内外，包括欧美、日韩、东南亚等国家和中国香港地区。

Li Shangfei is now the legal representative of Guizhou Dujuan Mingpinzhai tourism commodity development Co., Ltd. He has been engaged in the business of Duyun Maojian tea, national handicrafts and tourism commodities for more than 20 years. He is keen on protecting and developing local wild tea species. His products are sold at home and abroad, including Europe, America, Japan and South Korea, Southeast Asia, Hong Kong and other countries and regions.

四

毛尖远航

Tea industry itself has economic, cultural, ecological and social benefits, which is one of the most important economic industries in the world. Chen Zhongmao, academician of Chinese Academy of Engineering and leader of Chinese tea science, once said: "if the tea industry is compared to an aircraft, tea culture and tea technology are the two wings of the aircraft, effectively promoting and ensuring the take-off of the tea industry."

茶业本身具备经济效益、文化效益、生态效益、社会效益，是当今世界重要的经济产业之一。中国工程院院士、中国茶学学科带头人陈忠懋曾言："如果把茶产业比喻为一架飞机，茶文化和茶科技就是这架飞机的两翼，有力地推进和保障茶产业的起飞。"

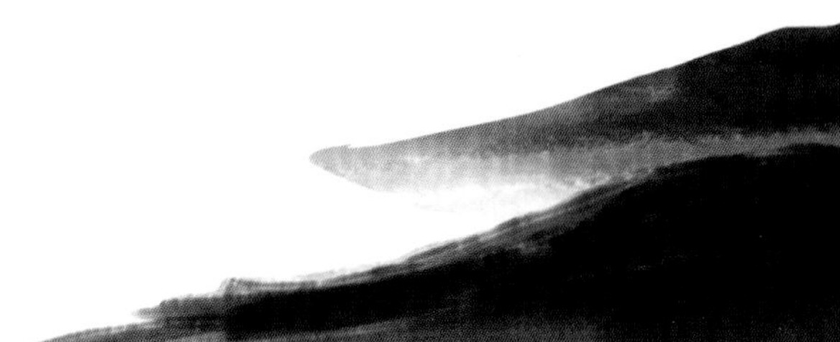

茶品清香（作者：傅宝）

经典 109

都匀毛尖辉煌历程

在都匀做茶叶生意的人都知道这样一个故事。1982年，都匀有两位师傅怀揣着亲手制作的两斤都匀毛尖茶，奔赴长沙参加评比。一个星期后，评审结果出来了。参加评比的86个茶品中，都匀毛尖名列前十。2016年以来，都匀毛尖连续4年入列中国茶叶品牌前20强。

People who do tea business in Duyun know such a story. In 1982, two masters from Duyun went to Changsha to participate in the contest with 1kg of Duyun Maojian tea made by themselves. A week later, the results of the review came out. Among the 350 tea products in the competition, Duyun Maojian ranked in the top 10. Since 2016, Duyun Maojian tea has been listed in the top 20 Chinese tea brands for 4 consecutive years.

经典 110

都匀毛尖引领黔茶出山

引领黔茶出山图（**作者：王殿维**）

贵州省委、省政府高度重视茶产业发展，把加快茶产业发展与生态文明建设、小康社会建设有机结合起来，制定了《贵州省茶产业提升三年行动计划》，将都匀毛尖列为全省主推的“三绿一红”黔茶品牌之首，赋予“引领黔茶出山”的重任。

Guizhou provincial party committee and government have attached great importance to the development of the tea industry, organically combined accelerating the development of tea industry with the construction of ecological civilization and well-off society, and formulated the Three-year Action Plan for the Promotion of the Tea Industry in Guizhou Province. Duyun was listed as the top brand of “three green and one red” Qian tea in the province, and the task of “leading Qian tea out of the mountain” was given.

寒香清韵（作者：王林）

经典 111

都匀毛尖跨越发展

2018 年黔南州涉茶企业（合作社）1300 余家，其中州级龙头企业 55 家，省级龙头企业 37 家，国家级龙头企业 1 家。通过多年的发展，都匀毛尖品牌已在全国设立专卖店 316 个、销售点 4728 个，入驻电商平台 358 个。

In 2018, there were more than 1300 tea related enterprises (cooperatives) in Qiannan Prefecture, including 55 state-level leading enterprises, 37 provincial-level leading enterprises and 1 national level leading enterprises. Through years of development, Duyun Maojian brand has set up 316 special stores, 4728 retail stores and 358 e-commerce platforms in China.

经典 112

都匀毛尖提质增速

紫气东来（作者：王林）

都匀毛尖韵味独具、历久弥香、顺势发展，是绿色茶、生态茶、健康茶、安全茶。黔南州 2018 年茶园种植面积已达 161.8 万亩，茶叶从业人员 38.8 万人，实现茶叶总产量 4.1 万吨，总产值 63.8 亿元。全州建成省级茶叶园区 5 个、州级茶叶园区 8 个，万亩以上乡镇 41 个、万亩专业村 23 个。

Duyun Maojian has a unique flavor, long lasting fragrance and favorable development, which is green tea, ecological tea, health tea and safety tea. In 2018, the tea plantation area of Qiannan prefecture has reached 1.618 million mu, with 388,000 tea practitioners. The total output of tea is 41,000 tons and the total output value is 6.38 billion yuan. There are 5 provincial tea parks, 8 state tea parks, 41towns with an area of more than 10,000 mu and 23 professional villages with an area of 10,000 mu.

集雅图（作者：王林）

经典 113

最具品牌经营力品牌

2019 年中国茶叶区域公用品牌价值评估结果揭晓，都匀毛尖以 32.90 亿元人民币的公用品牌价值，位列榜单第 11 位，被评选为中国“最具品牌经营力品牌”。

In 2019, the evaluation results of public brand value in China's tea region were announced. Duyun Maojian ranked 11th in the list with a public brand value of 3.290 billion yuan, and was selected as the "most powerful brand in brand management" in China.

中国品牌农业神农奖

富贵平安（作者：王林）

2019 年神农氏诞辰日（农历四月廿六日），在国家首届“品牌农业神农论坛”上，都匀毛尖茶作为在品牌农业建设中作出突出贡献和积极探索的区域公共品牌和企业品牌，被授予“2019（首届）中国品牌农业神农奖”。

On the birthday of Shennong in 2019 (April 26, the lunar calendar), at the first national "Brand Agriculture Shennong Forum", Duyun Maojian tea was awarded the "2019 (first) China Brand Agricultural Shennong Award" as a regional public brand and enterprise brand that made outstanding contributions and actively explored in brand agricultural construction.

经典 115

中国高端绿茶特色品牌

都匀毛尖茶的好品质源于生态优良的环境、绿色健康的加工。从好原料变成好产品，绿色生态贯穿始终。在 2019 都匀毛尖(国际)茶人会开幕式上，都匀毛尖荣获“中国高端绿茶特色品牌”。

The good quality of Duyun Maojian tea comes from the excellent ecological environment and green and healthy processing. From good raw materials to good products, green ecology runs through the whole process. At the opening ceremony of Duyun Maojian (International) Tea People's Conference in 2019, Duyun Maojian won the title of "China's high-end green tea brand".

研经图（作者：贺成才）

敬茶歌（作者：范德昌）

经典 116

互联网催生毛尖茶产业

都匀市政府不断探索以茶产业为核心的农业产业融合发展格局。基于“互联网 +”的大背景下，建立了都匀毛尖茶商城，先后引进 235 家茶企入驻。启动建设了 50 个农村电商网店，实现了“线上”与“线下”共同发展。

Duyun municipal government has been exploring the integration development pattern of agricultural industry with tea industry as the core. Maojian tea shopping mall was established in Duyun based on the “Internet plus” background, and 235 tea enterprises have been introduced to settle down. 50 rural e-commerce online stores have been launched and realized the common development of “online” and “offline”.

茶品人生味（作者：范德昌）

经典 117

都匀毛尖与文旅结合的新亮点

螺蛳壳茶产区地处都匀市“百里毛尖长廊”产业带，海拔 1480 米，远离城市工业区，青山绿水相环绕，云雾婆娑出美景。基于优良的生态环境，都匀市政府依托螺丝壳茶产区，发展茶旅结合的特色旅游业，全力打造以旅游者“参观螺蛳壳风景区 -- 体验采茶、制茶过程——品尝鉴赏都匀毛尖茶”的旅游新亮点。

Luosike tea production area is located in Duyun “hundred-mile Maojian corridor” industrial belt, about 1480 meters above sea level, away from the urban industrial area, surrounded by green mountains and rivers, with beautiful scenery. Based on the excellent ecological environment, the Duyun municipal government, relying on the Luosike tea production area, develops the characteristic tourism of tea tourism combination, and makes every effort to create a new tourist highlight of “visiting Luosike scenic spot – experiencing tea picking, tea making process – tasting and appreciating Duyun Maojian tea”.

品茗图（作者：尚秋香）

经典 118

都匀毛尖拓展美好茶生活

都匀毛尖茶是文化名茶，努力打造“茶园变公园、茶区变景区、茶人变茶师、茶山变金山、茶旅助康养、茶乡奔小康”的茶旅休闲养生一体化综合体，致力于为中国乃至世界茶人拓展全新的美好茶生活、茶空间、茶天地。

Duyun Maojian tea is a famous cultural tea, striving to build a tea tourism leisure and health integration complex that " tea garden changes into Park, tea area changes into scenic spot, tea people changes into tea master, tea mountain changes into golden mountain, tea tourim helps to maintain health, and tea country goes to a well-off society", so as to expand a new and better tea life, tea space and tea world for tea people in China and even in the world.

扬帆远航　再创佳绩

福寿康宁（作者：子晨）

2019 年在第三届中国国际茶叶博览会上，都匀毛尖茶清幽甘醇的茶香受到茶商们的追捧和好评，5 天内共达成意向性订单近 2000 万元。州政府领导在肯定成绩的同时，要求："参展企业在推广茶产品的同时，还应优化产品包装，要设计出更加符合展会需求的旅游包装，用精美小巧、性价比高的产品吸引更多消费者和客商。"

In 2019, at the third China International Tea Expo, the tea fragrance of Duyun Maojian was sought after and praised by tea merchants. In five days, an intentional order of nearly 20 million yuan was reached. While affirming the achievements, the leader of the state government required: "while promoting tea products, exhibitors should also optimize product packaging, design tourism packaging more in line with the needs of the exhibition, and attract more consumers and merchants with exquisite, small and cost-effective products."

Chapter V
Tea Painting Forum

Tea painting, with its unique artistic charm in Chinese tea culture, is favored by the majority of tea people. From the expression way, it belongs to the traditional Chinese ink painting, which mainly takes landscape and characters as the content, and integrates into tea drinking activities. This kind of tea painting art with national and regional characteristics has attracted more and more attention. Many researchers who like the art of tea painting begin to study it deeply, in order to find more abundant ideological connotation and aesthetic value.

第五篇

茶画春秋

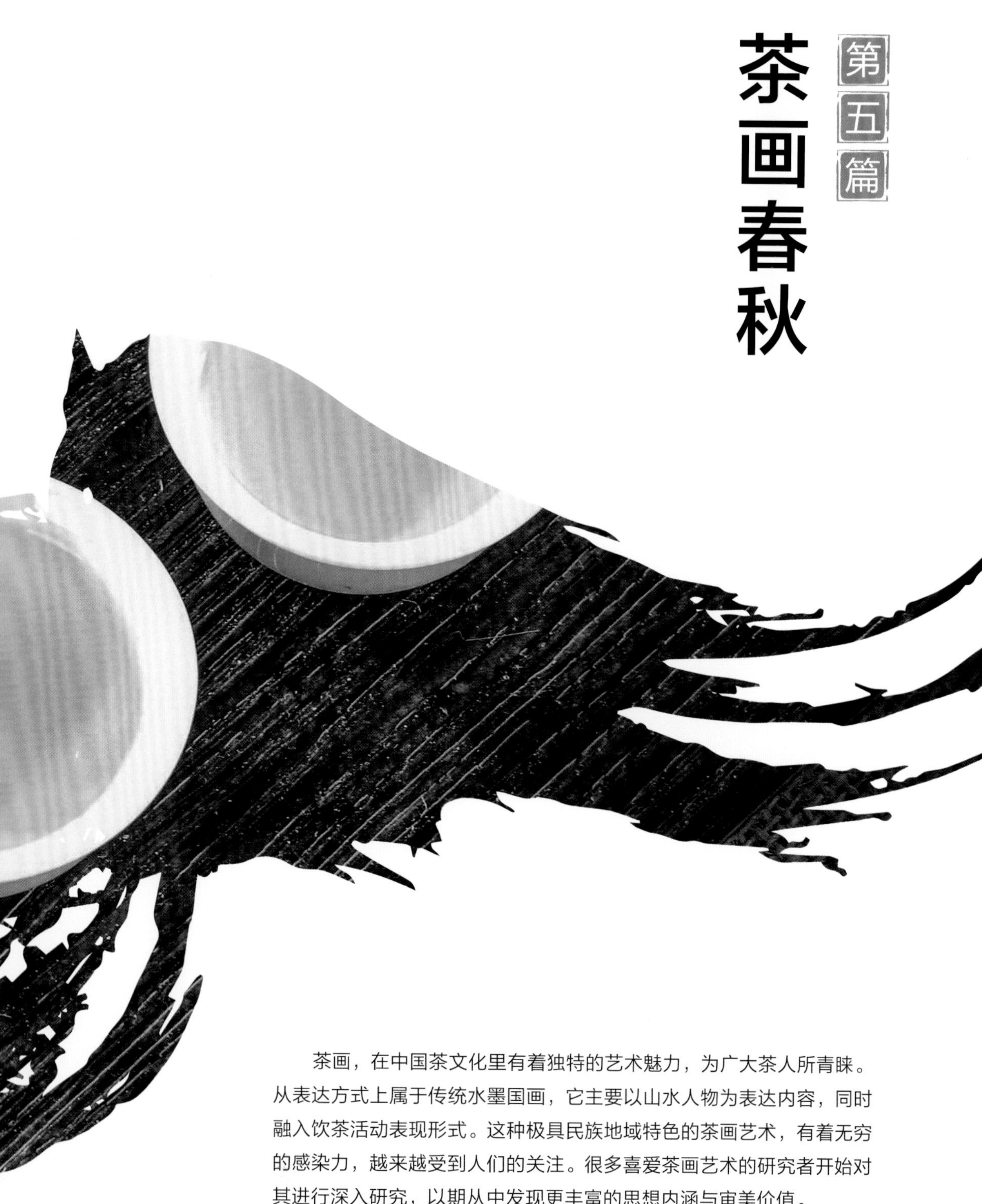

茶画，在中国茶文化里有着独特的艺术魅力，为广大茶人所青睐。从表达方式上属于传统水墨国画，它主要以山水人物为表达内容，同时融入饮茶活动表现形式。这种极具民族地域特色的茶画艺术，有着无穷的感染力，越来越受到人们的关注。很多喜爱茶画艺术的研究者开始对其进行深入研究，以期从中发现更丰富的思想内涵与审美价值。

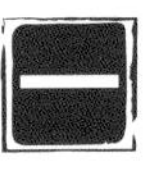

名家说茶

The ups and downs of every piece of tea are a kind of fate, no emptiness and no annihilation. Since ancient times, many famous people have made an indissoluble bond with tea. They have not only written many poems praising tea, but also left many interesting anecdotes about tea making and tea tasting. The fate of tea is closely connected with that of the country.

每一片茶叶的浮沉，都是一种缘定，不空不昧。自古至今，许多名人大家与茶结缘，不仅为茶写下了众多吟咏称道的诗章，还留下了不少煮茶品茗的趣事逸闻。茶运与国运紧密相连。

乾隆茶语

国不可一日无君，君不可一日无茶。

A country cannot be without a king for a day, and a king cannot be without tea for a day.

——Qian Long

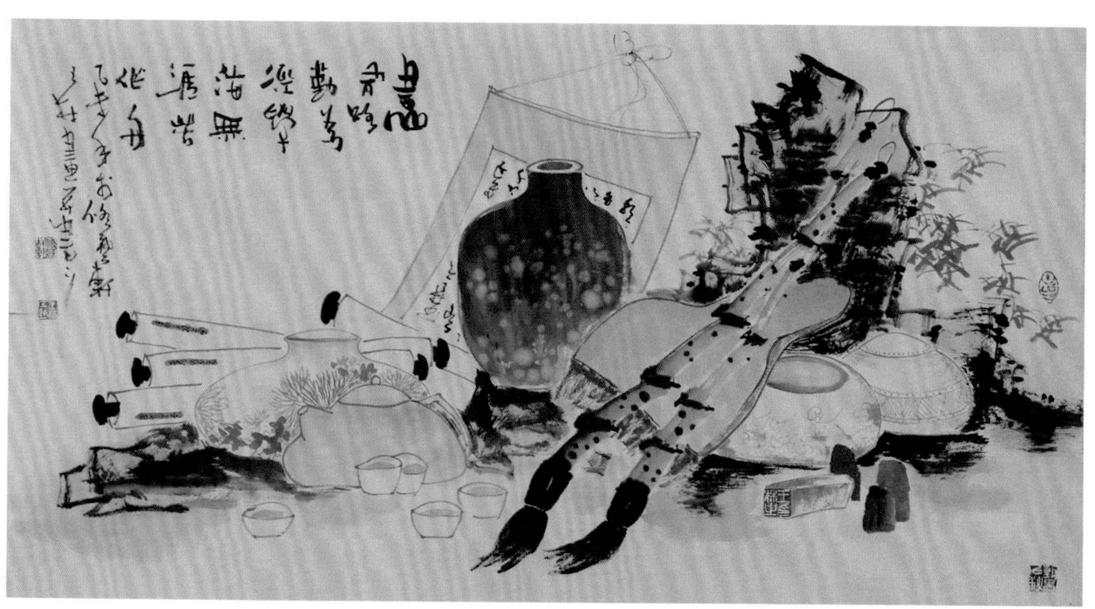

清供图（作者：王林）

鲁迅茶语

我是读书人（作者：吴梦千）

有好茶喝，会喝好茶，是一种“清福”。

Having good tea to drink, and being able to drink good tea, is a kind of pure happiness.

——Lu Xun

静夜水边独坐（作者：李国锰）

经典 122 老舍茶语

有一杯好茶，我便能万物静观皆自得。烟酒虽然也是我的好友，但它们都是男性的——粗莽、热烈、有思想，可也有火气。未若茶之温柔，雅洁，轻轻地刺激，淡淡地相依，是女性的。

With a good cup of tea, I can enjoy everything. Although tobacco and wine are my good friends, they are all male — rough, warm, thoughtful, but also passion. Unlike tea, which is gentle, elegant and clean, gently stimulated, and lightly dependent, like female.

——Lao She

闲趣（作者：李国锰）

经典 123

周作人茶语

喝茶当于瓦屋纸窗之下，清泉绿茶，用素雅的陶瓷茶具，同二三人共饮，得半日之闲，可抵十年的尘梦。喝茶之后，再去继续修各人的胜业，无论为名为利，都无不可，但偶然的片刻悠闲断不可少。

Tea should be tasted under the paper window of a tile house. Drink with two or three people with clear spring, green tea and elegant ceramic tea sets. Half a day's leisure can be worth ten years dream. After tea, you can continue to build your own business. No matter for reputation or benefit, it's all necessary. But occasional monents of leisure are indispensable.

——Zhou Zuoren

经典 124

三毛茶语

第一道苦若生命，第二道甜似爱情，第三道淡如微风。

The first tea is as bitter as life, the second as sweet as love, and the third as light as breeze.

—— San Mao

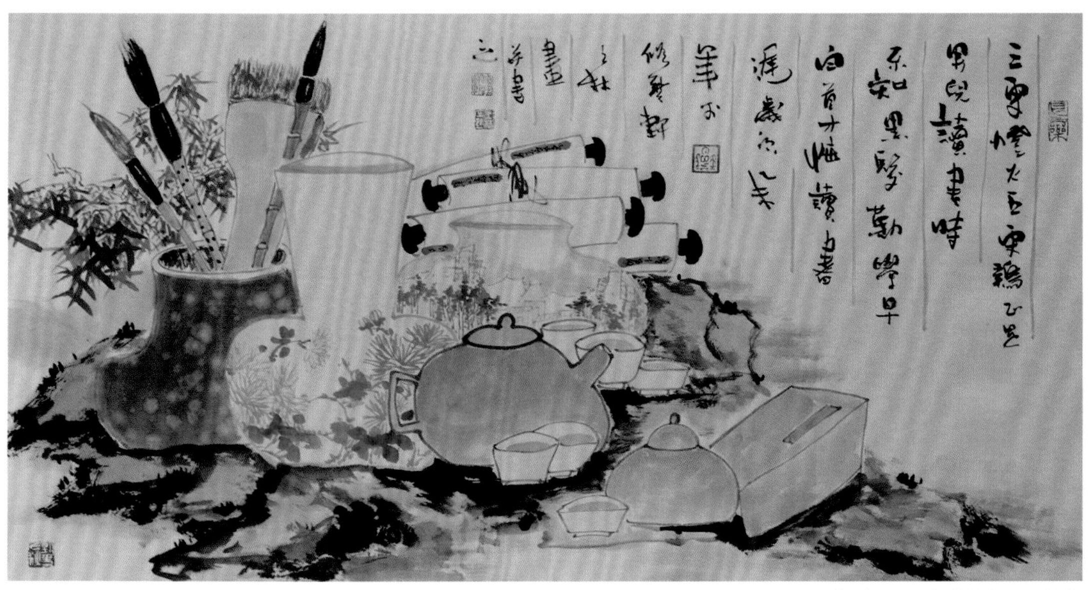

三更灯火五更鸡（作者：王林）

清品图（作者：春善）

杨绛茶语

记不起哪一位英国作家说过，“文艺女神带着酒味”“茶只能产生散文”。而咱们中国诗，酒味茶香，兼而有之，“诗清只为饮茶多”。也许这点苦涩，正是茶中诗味。

I can't remember a British writer who said, "The goddess of literature and art has a taste of wine", "Tea can only produce prose". But our Chinese poetry, the wine taste tea fragrance, has both. "The poem is elegant only because of drinking more tea." Maybe this bitter taste is just poetry in tea.

——Yang Jiang

去岁残荷尚
在今年新花又开
多少人间事情
怎能重新再来

残荷新花（作者：李国锰）

经典 126
林语堂茶语

严格地说起来，茶在第二泡时为最妙。第一泡譬如一个十二三岁的幼女，第二泡为年龄恰当的十六岁女郎，而第三泡则是少妇了。

Strictly speaking, tea is the best at the second time. The first tea is like a little girl aged 12 or 13, the second is a 16-year-old girl with the right age, and the third is a young woman.

——Lin Yutang

问道茗香

The ancients said: "tea is the elixir of health preservation and the wonderful skill of prolonging age." Su Dongpo of the Northern Song Dynasty also said: "why do you have to take a pill of emperor Wei? Just drink Lu Tong's seven bowls of tea." Ancient life practice and modern scientific verification agree that tea is a healthy drink, and drinking tea is beneficial to physical and mental health.

古人云："茶乃养生之仙药，延龄之妙术。"北宋苏东坡也说："何须魏帝一丸药，且尽卢仝七碗茶。"古代生活实践和现代科学验证一致认为，茶是健康饮品，饮茶有利身心健康。

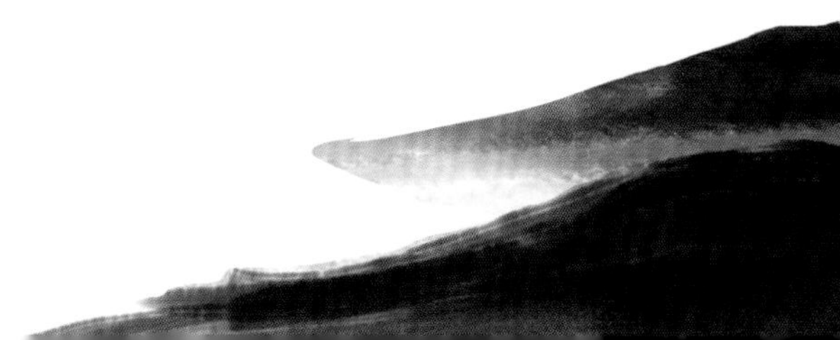

东坡煎茶图（作者：李彦岭）

茶，发于神农，闻于鲁周公，兴于唐朝，盛于宋代。茶文化糅合了儒、道、佛诸派思想，是中国文化中的一朵奇葩。中国人讲究喝茶之道，历来有客来敬茶的传统礼节。注重春饮花茶，夏饮绿茶，秋饮青茶，冬饮红茶。

Tea, discovered by Shennong, was famous in the period of Lu Zhougong. It rose in Tang Dynasty and flourished in Song Dynasty. Tea culture is a woderful flower of Chinese culture, which combines Confucianism, Taoism and Buddhism. Chinese people pay attention to the way of drinking tea, and there have always been traditional etiquette for guests to serve tea. Pay attention to drinking jasmine tea in spring, green tea in summer, oolong tea in autumn, black tea in winter.

茶之六度图并记（作者：范德昌）

上古时期，人们把茶当成长生不老的仙药。《神农本草》中记载：“神农尝百草，日遇七十二毒，得茶而解。”唐代大医学家陈藏器在《本草拾遗》里也说过：其他药都是各个病之药，茶为“万病之药”。可见古人对茶叶功效的重视程度。

In ancient times, people regarded tea as an elixir for growing up. Shennog Materia Medica recorded that: “Shennong tasted all kinds of herbs, and met seventy-two poisons every day, but got rid of poisons by tea.” Chen Cangqi, a great medical expert in Tang Dynasty, also said in the book “the Collection of Materia Medica”: other medicines are all medicines for various diseases, and tea is the medicine for all kinds of diseases.So wen can see that how important the ancients attached to the tea.

茶香醉人（作者：白林）

经典 129

减缓衰老　幸好有茶

自由基的破坏活动是人体衰老的罪魁祸首。研究表明，茶是天然植物中，清除自由基能力最强的一种。其茶多酚的抗氧性明显优于维生素 E，抗衰老效果要比维生素 E 强 18 倍，且与维生素 C、E 有增效效应。

The destruction of free radicals is the main cause of human aging. The research shows that tea is one of the natural plants with the strongest ability of scavenging free radicals. The antioxidant activity of tea polyphenols was significantly better than that of vitamin E. The anti-aging effect was 18 times stronger than that of vitamin E, and it had synergistic effect with vitamin C and E.

心经（作者：周子杰）

经典 130

广岛现象

据《科学养生》1997 年 03 期文章介绍，1945 年 8 月，日本广岛原子弹轰炸致使 10 多万人丧生，数十万人遭受辐射伤害。若干年后，大多数人患上白血病或其他各种肿瘤，先后死亡。但研究却发现有三种人侥幸无恙：茶农、茶商、茶癖者，这一现象被称为“广岛现象”。

According to the introduction of Scientific Health Preservation(issue 03, 1997), in August 1945, the atomic bombing in Hiroshima in Japan killed more than 100,000 people. Hundreds of thousands of people suffered radiation damage. Several years later, most people suffered from leukemia or other kinds of tumors, and died successively. But the study found that there were three kinds of people who were lucky: tea farmers, tea merchants and tea addicts. This phenomenon is called “Hiroshima phenomenon”.

读书图（作者：文飞扬）

经典 131

喝茶降低胆固醇

据研究显示，不管是喝红茶，还是喝绿茶，都有助于降低胆固醇，减少由吸烟引起的细胞伤害，甚至对心脏、前列腺等器官都起到强健保护作用。国外研究人员在癌细胞的培养液中加入红茶、绿茶或苏打后发现，癌细胞生长速度显著减缓。

According to research, whether you drink black tea or green tea, can help poeple lowering cholesterol, even protect our heart and prostate. Foreign researchers found that the growth rate of cancer cells slowed down significantly after adding black tea, green tea or soda to the culture medium of cancer cells.

绿茶降火

抚琴图（作者：贺成才）

绿茶是名茶最多、研究最广泛的茶类。儿茶素等多酚类化合物被公认为是绿茶中对健康有益的关键成分，具有多种保健功能，包括预防癌症、改善心血管健康、减肥、抵御电离辐射等。此外，绿茶还是维生素种类最多、含量最丰富的茶类。与其他茶类相比，绿茶的抗癌功效更好。多项研究表明，绿茶能降低乳腺、前列腺、肺、口腔、膀胱、结肠、胃、胰腺等多部位肿瘤发生的危险性。

Green tea is the most famous and widely studied tea. Catechins and other polyphenols are recognized as the key components of green tea, which are beneficial to health. They have many health functions, including cancer prevention, cardiovascular health improvement, lossing weight, and resistance to ionizing radiation, etc. In addition, green tea is the tea with the most kinds of vitamins and the most abundant content. Compared with other teas, green tea has better anticancer effect. A number of studies have shown that green tea can reduce the risk of tumors in breast, prostate, lung, mouth, bladder, colon, stomach, pancreas and other parts.

人生如茶图并记（**作者：范德昌**）

经典 133

喝绿茶可防晒

美国发现饮用绿茶可达防晒效果。近期美国有一项研究指出，绿茶中的儿茶素有很强的抗氧化功能，喝绿茶等于将含有儿茶素的护肤品涂抹在皮肤上，即使被猛烈阳光照射，也可让导致皮肤晒伤、松弛和粗糙的过氧化物减少约 1/3。

Green tea has been found to protect against sun in the United States. A recent study in the United States pointed that catechins in green tea have strong antioxidant function. Drinking green tea is equivalent to applying skin care products containing catechins on the skin. Even if it is exposed to strong sunlight, it can reduce the peroxides that cause skin sunburn, relaxation and roughness by about 1/3.

品茶图（作者：何奕勇）

经典 134

青茶润燥减肥

乌龙茶又名青茶，既有绿茶的清香，又有红茶醇厚的滋味。乌龙茶属于半发酵茶类，加工工艺介于绿茶和红茶之间。与其他茶类相比，乌龙茶在减肥方面效果较好。从中医的角度来说，乌龙茶性平，能清除体内积热，特别适合秋天饮用，可缓解秋燥。

Oolong tea, also known as Cyan tea, has the fragrance of green tea and mellow taste of black tea. Oolong tea belongs to semi-fermented tea, and its processing technology is between green tea and black tea. Compared with other kinds of tea, oolong tea has a better effect in reducing weight. From the perspective of traditional Chinese medicine, It is especially suitable for drinking in autumn and can alleviate the dryness in autumn.

茶道图并记（**作者：范德昌**）

经典135 白茶抑菌抗辐射

白茶的加工工艺比较简单，其成茶满披白毫，芽叶连梗，形态自然素雅，色泽银白灰绿，汤色清淡。白茶的化学成分一般与绿茶比较接近。与其他茶类相比，白茶的抗菌效果比较好。此外，白茶也具有较好的抗辐射效果，在美国和欧洲地区白茶提取物被用于脸部护肤品的开发。

The processing technology of white tea is relatively simple. This kind of tea is covered with pekoe. The buds and leaves are connected with peduncles, and the shape is natural and elegant. The color is silver, gray and green and the soup is light. The chemical composition of white tea is generally close to that of green tea. Compared with other teas, white tea has better antibacterial effect. In addition, white tea also has a good radiation resistance effect. In the United States and Europe, white tea extract is used in the development of facial skin care products.

红茶暖胃护心

红尘渺渺（作者：李国锰）

红茶属于全发酵茶，在加工过程中，茶多酚氧化成茶黄素和茶红素，多数糖类物质水解成可溶性糖，从而产生红茶特有的香气和口感。茶黄素是红茶中最主要的功能性成分。饮用红茶有助调节人体动脉中低密度脂蛋白和高密度脂蛋白的含量，从而降低心血管疾病的发生概率。此外，茶黄素还可抗氧化、抗癌、预防慢性炎症和肥胖。与其他茶类相比，一般红茶预防心血管疾病的功效较好。从中医的角度来说，红茶性温，有暖胃的作用，虚寒体质者和老年人宜饮性温的红茶。

Black tea belongs to fully fermented tea. In the process of processing, tea polyphenols are oxidized into theaflavins and thearubigins, and most sugar substances are hydrolyzed into soluble sugar, thus producing the unique aroma and taste of black tea. Compared with other teas, black tea has a better effect on cardiovascular disease prevention. From the perspective of traditional Chinese medicine, black tea has the function of warming the stomach. People with deficiency cold constitution and the elderly should drink black tea with warm nature.

经典 137

黑茶御寒降脂

黑茶是中国特有的茶类，生产历史比较悠久，种类比较丰富，例如云南普洱茶、湖南茯砖茶等。黑茶属于后发酵茶，茶叶在微生物作用下会发生一系列复杂的化学反应，生成一些对人体有益的功能性成分。黑茶具有降血糖、降血脂、抗病毒等保健功能，降脂作用可能与其中的他汀类化合物有关。此外，黑茶中的矿物质种类丰富，且有很多种含量高于其他茶类。黑茶性温，有助御寒，适合虚寒体质者。

Dark tea is a special kind of tea in China, which has a long history of production and rich varieties, such as Yunnan Pu'er tea, Hunan Fuzhuan tea, etc. Dark tea belongs to the post fermented tea. Under the action of microorganism, tea will produce a series of complex chemical reactions, and produce some functional components beneficial to human body. Dark tea has the health care functions of lowering blood sugar, blood lipid and antiviral. The effect of lowering blood lipid may be related to the statins. In addition, there are many kinds of minerals in dark tea, and many of them are higher than other teas. Dark tea is warm in nature, which helps keep out cold. It's suitable for people with deficiency cold constitution.

赏荷图（作者：贺成才）

书中乾坤大（作者：傅宝）

经典 138

茶可提神助眠

茶同时具有提神和养神两方面的作用。茶叶刚泡开前 3 分钟左右，大部分是咖啡碱溶入茶汤中，这时的茶就具有明显的提神醒脑作用。而再往后，茶叶中的鞣酸逐渐溶解到茶水中，抵消了咖啡碱的作用，就不容易再使人有明显的生理上的兴奋了。所以，只要把一开始冲泡约 3 分钟的茶水倒掉，再续上开水冲泡饮用，就会起到养神的作用。

Tea has both refreshing and nourishing functions. About 3 minutes before the tea leaves are brewed, most of them are caffeine dissolved in the tea soup. At this time, the tea has an obvious refreshing effect. Later, the tannic acid in the tea gradually dissolves into the tea water, counteracting the effect of caffeine, so it is not easy to make people have obvious physiological excitement. So, just pour out the tea that is brewed for 3 minutes at the begining, and then add boiling water for brewing and drinking, it will play a role of nourishing the mind.

渴饮清泉闷煮茶（**作者：杨平**）

经典 139

茶水洗头染出健康色

洗过头发之后再用茶水洗涤，可使头发乌黑柔软，富有光泽。用很浓的红茶洗头，就能让头发变成棕色或黑色。如果用很浓的丁香茶洗头，头发就会变成红色，而且还没有毒性。即使不想染色，洗过头发后用废茶水冲洗，也能去垢除腻，让头发乌黑、柔软有光泽。

After washing the hair, wash it with tea water, which can make the hair dark, soft and lustrous. Wash your hair with strong black tea, and it will turn brown or black. If you wash your hair with a strong clove tea, it will turn red and not toxic. Even if you do not want to dye your hair, wash it with waste tea water after washing it, you can get rid of dirt and greasiness, making it dark, soft and shiny.

好吃不过茶煮饭

《本草拾遗》记载，用茶水煮饭“久食令人瘦”。方法是取适量茶叶加水冲泡，待茶叶泡开后，滤去茶叶取汤煮饭。茶叶的清香融入米饭的香甜，煮好的米饭不仅色、香、味俱佳，而且具有诸多保健功能。茶煮饭已成为爱茶人士的“食”茶新时尚。

According to the Collection of Materia Medica, cooking rice with tea is “thin after a long time of eating”. The method is to take a proper amount of tea and add water for brewing. After the tea is brewed, fiter the tea leaves and take the soup for cooking. The fragrance of tea blends into the sweetness of rice. The cooked rice not only has good color, fragrance and taste, but also has many health functions. Cooking rice with tea has become a new fashion for the tea lovers.

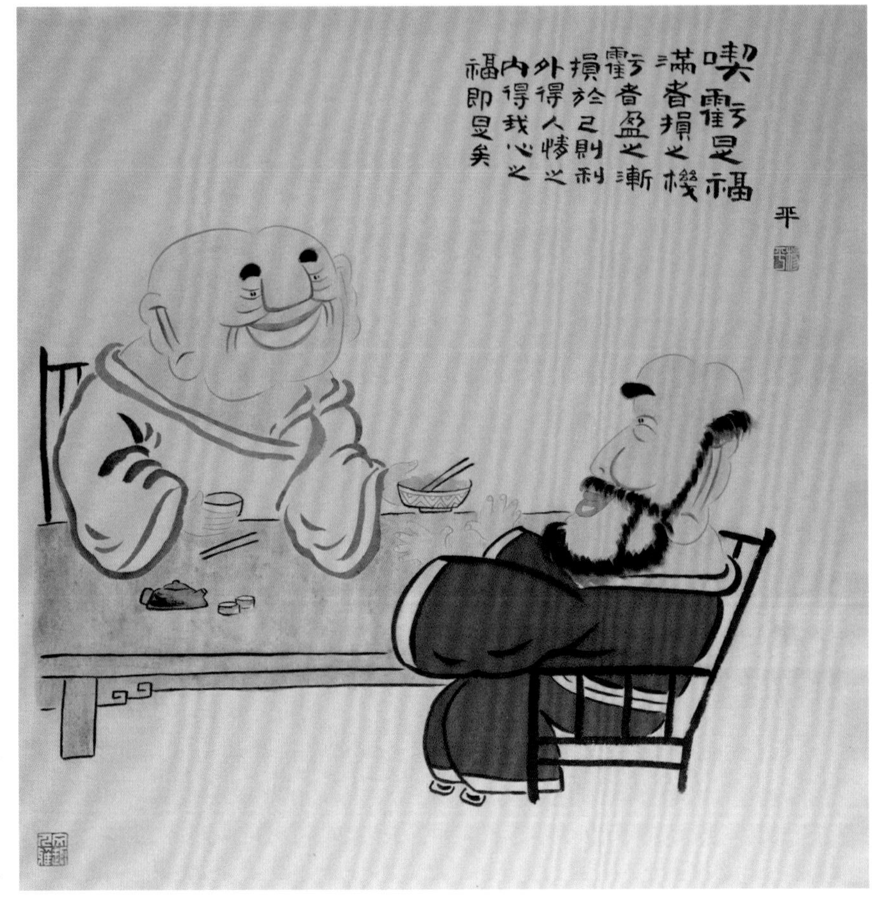

吃亏是福（作者：杨平）

经典 141

酒后不宜饮茶

心悟（作者：洪君平）

现代医学证实，酒后饮茶，特别是饮浓茶，会对肾脏造成不良影响。因此，想在喝多之后解酒，可进食些柑橘、梨、苹果之类水果，喝西瓜汁更好。如无水果，冲杯糖水也能帮助解酒。

Modern medicine has proved that drinking tea after drinking alcohol, especially strong tea, can cause adverse effects on the kidney. Therefore, if you want to sober up after drinking more, you can eat some citrus, pear, apple and other fruits, and drink watermelon juice better. If there is no fruit, a glass of sugar water can also help sober up.

经典 142
每日饮茶多少为宜

煮茶图（作者：杨平）

一天喝几杯茶为宜？曾有人戏称：一杯为品，两杯为饮，三杯四杯就是饮牛饮马的蠢物了。其实，习惯于饮茶的人，可以在上午下午各喝两至三杯。不仅可以喝茶水，喝剩的茶叶也可以吃，其中富含维生素 E、食物纤维等有助健康的物质。

How many cups of tea a day is appropriate? Someone once joked that: one cup is for tasting, two cups for drinking, three cups and four cups are fools who drink cattle and horses. In fact, people who are used to drinking tea can drink two to three cups each in the morning and afternoon. It can not only drink tea, but also eat the leftover tea, which is rich in vitamin E, food fiber and other health promoting substances.

喝好眼前这壶茶（作者：杨平）

经典 143

上善若水

“上善若水，水善利万物而不争”。《茶经》中记载：“山水上，江水中，井水下。”对煮水的要求，《茶经》曰：“其沸如鱼目，微有声，为一沸。边缘如涌泉连珠，为二沸。腾波鼓浪，为三沸。已上，水老，不可食也。”当好水遇见好茶，本身就是一种幸运。

“The best is like water. Water is good for all things without dispute.” According to the Book of Tea, “the water on the mountain is superior, the water in the river is medium, and the water in the well is inferior.” As for the requirement of boiling water, the Book of Tea said: “it boils like the eyes of fish, with a slight sound, which is called first boiling. The edge is like water beads out of spring, second boiling. The water billows, which is third boiling. Keep boiling, the water is old and cannot be eaten.” When good water meets good tea, it is a kind of luck in itself.

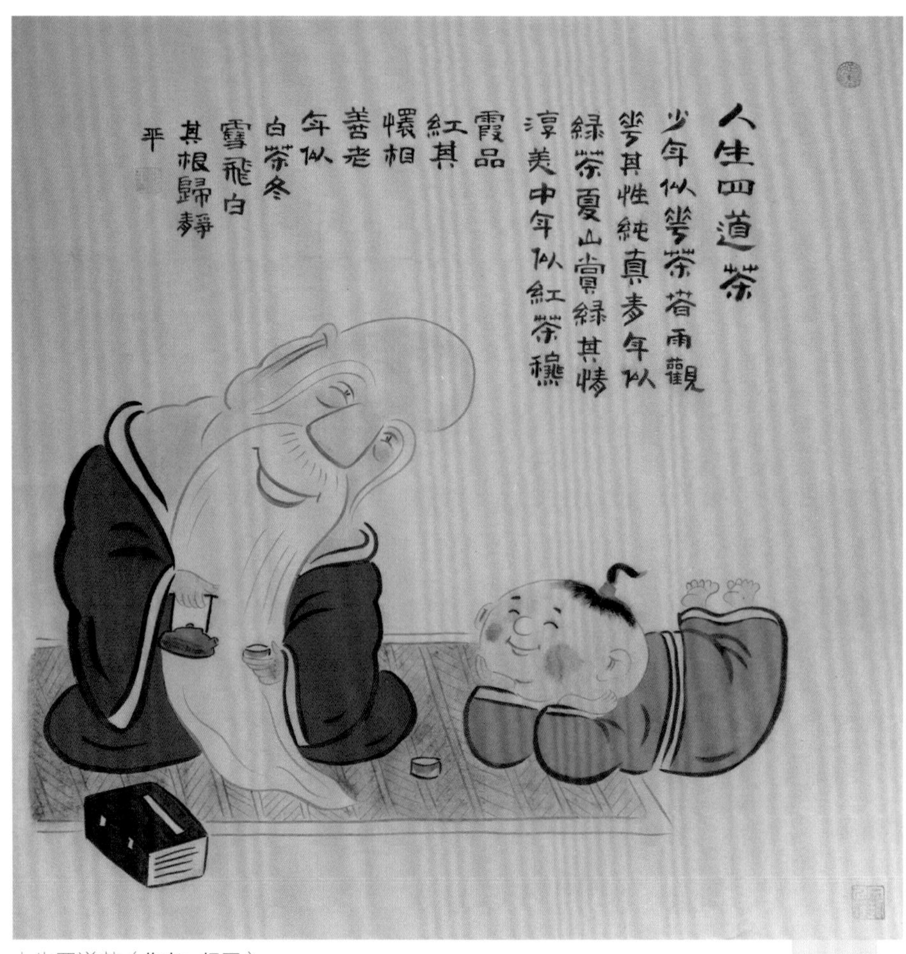

人生四道茶（作者：杨平）

经典144

香靠冲　汤靠吊

“香靠冲，汤靠吊”的意思是说，如果想要茶汤高香，就快水猛冲，让茶叶在容器中翻腾激荡，和水充分摩擦。如果想要茶汤绵密柔软，就让水流在一个点上稳定而缓慢地注入泡茶器皿。

“Fragrance depends on rush, soup depends on hanging”. That is to say, if we want the tea soup to be highly fragrant, we will rush the water quickly, let the tea toss and stir in the container, and thoroughly grind with water. If we want the tea soup to be sticky dense and soft, let the water flow steadily and slowly into the tea making vessel at a point.

经典 145

定点注水

有些茶因其本身内涵物质丰富，咖啡碱含量高，汤质厚重感强，冲泡时要求沿杯壁，定点注水，避免击打茶叶，且不宜出汤过急。这样做的目的是，一方面可以获得稠厚的汤感，另一方面可以避免咖啡碱过度析出，减轻茶汤的苦涩。

Some teas are rich in substance and caffeine, so the soup is thick and strong. Therefore, when brewing, it is required to inject water along the wall of the cup at fixed point to avoid hitting tea leaves, and it is not suitable to rush out the soup. The purpose of this is to obtain a thick soup feeling on the one hand, on the other hand to avoid excessive caffeine precipitation, so as to reduce the bitterness of tea soup.

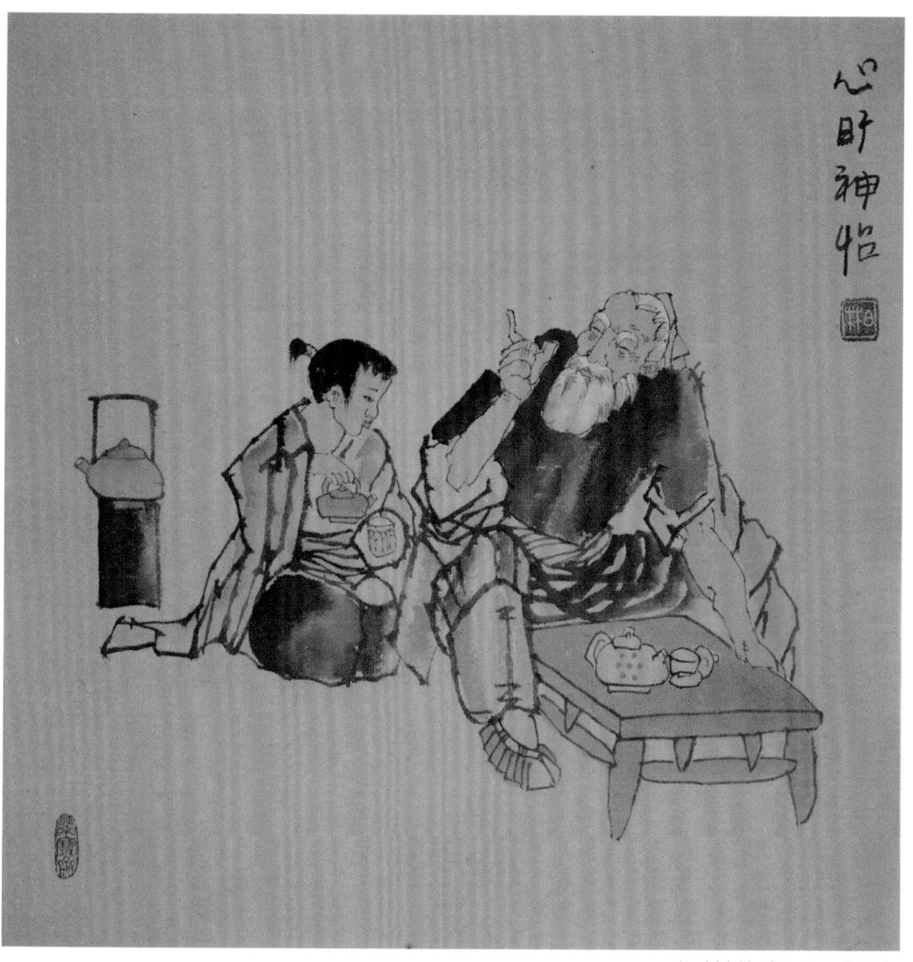

心旷神怡（作者：白林）

水线

从来佳茗似佳人（作者：杨平）

所谓“水线”，其实就是指注水方式。注水方式的不同，影响着茶汤口感的不同。注水方式常见的有四种：高冲、高吊、低冲、低吊，注水点还分螺旋形注水、环圈注水、单边定点注水、正中定点注水等。

The so-called "water line" actually refers to the way of water injection. Different ways of water injection affect the taste of tea soup. There are four common ways of water injection: high rush, high hanging, low rush, low hanging. Water injection points are also divided into spiral water injection, ring water injection, single side fixed-point water injection, central fixed-point water injection, etc.

慈母手中线（作者：白林）

经典 147

绿茶保存法

保存绿茶时要尽量装满，减少贮存空间内的空气。可用干燥陶罐、纸罐、锡罐、马口铁罐等存放茶叶，罐内底部放置双层棉纸，罐口放置两层棉布而后压上盖子。最好能预备一台专门贮存绿茶的小型冰箱，设定温度在 -5℃ 以下。也可以将茶叶贮存在一般冰箱的冷冻库，其内不能再贮存其他东西，以防串味。还可以将拆封的茶叶倒入清洁干燥的热水瓶内，塞紧塞子存放。

Keep green tea as full as possible to reduce the air in the storage space. Tea can be stored in dry pottery can, paper can, tin can and tinplate can, etc. Two layers of cotton paper are placed at the bottom of the can, two layers of cotton cloth are placed at the can mouth, and then the lid is pressed on. It is better to prepare a small refrigerator for storing green tea, with the temperature set below -5℃. The unsealed tea can also be poured into a clean and dry water hot bottle, which be corked up for storage.

茶来茶去

It is not easy to make a cup of tea in our life. Because the style of tea is rich and complex, not only reflected in the production process, many of them need years of precipitation. Just as tea is not made in a day, you should brew it once, twice, and three times. Such a tea flavor can come out, and the aftertaste can last for a long time.

人生一世，草木一秋，生活中要成就一杯茶是不容易的。因为茶的风格丰富繁杂，不仅体现在制作工艺上，很多都需要岁月的积淀。正如茶非一日而成，泡茶也要一泡、两泡、三泡……如此茶味才出得来，回味才历久弥香。

梅花牡丹茶图（**作者：宋东海**）

经典 148 茶字演变

民俗专家宋国强说，茶最早的表述文字是“查”，意为用餐后喝一碗茶汤，以减除油腻、消化食物、排除毒素。后到汉代为“荼”，直到唐代，茶成了人们的日常饮料，使用频率越来越高，便把“荼”字减去一横，成为今日的“茶”字了。

Song Guoqiang, a folklore expert, said that the earliest expression of tea was “ 查 ”, which means drinking a bowl of tea soup after eating to reduce greasiness, digest food and eliminate toxins. Later, It was called“ 荼 ”in the Han Dynasty. Until the Tang Dynasty, tea became people's daily drink, more and more frequently used, so the word “ 荼 ” was changed into today's word “ 茶 ”(tea).

茶言观色（作者：心海）

经典 149 荼茶之别

陆羽之前，“荼”“茶”不分，一个“荼”字，两家共用。一种是草本植物，开黄花，味苦可作蔬菜的称“荼”。再一种是木本植物，叶作饮料的“槚”，也称作“荼”。名同实异，难免混淆不清。陆羽先生在著《茶经》时，把“荼”字减去一画作“茶”，其价值远远超出了文字学的范畴，在自然科学史上也有它重要的意义。

Before Lu Yu, there was no distinction between the two Chinese words “ 荼 ” and “ 茶 ”. One is herbaceous plant, with yellow flowers and bitter taste, which can be used as a vegetable. Another is woody plant, whose leaves can be used as drinks. Two kinds of plants have the same name, which is easy to be confused. When Mr. Lu Yu wrote the Book of Tea, he changed “ 荼 ” into “ 茶 ”(tea). Its value goes far beyond the scope of literature, and it also has its important significance in the history of natural science.

经典 150

茶树可贵

茗思图（作者：贺成才）

茶树为深根木本植物，是一种良好的绿色生态保护树种，具有较强的水土保持能力。常言道：“一年种二年管，三年见效五年赚。一次投入细管护，多年受益能致富。”是实现“绿水青山”“金山银山”的重要载体。

Tea is a kind of deep rooted woody plant, which is a good green ecological protection tree species with strong soil and water conservation capacity. As the saying goes: “we plant tea trees in the first year and manage them in two years. We can see results in the third year and make money in the fifth year. If you invest in it once and take care of it carefully, you will become rich after years of benefit.” It is an important carrier to realize “green water green mountain” and “golden mountains and silver mountains”.

清品图（作者：王林）

经典 151

茶的八种别称

茶在古诗文中曾有荼、茗、不夜侯、消毒臣、涤烦子、清风使、余甘氏、清友八种别称。其中不夜侯显然道明茶有醒脑、破睡之功效。消毒臣说明茶可以消酒肉毒。涤烦子可洗去心中烦闷，有诗云：“茶为涤烦子，酒为忘忧君。”清风使乃从卢仝的茶歌中而来。世称橄榄为余甘子，茶为余甘氏。唐代姚合品茶诗云：“竹里延清友，迎风坐夕阳。”

In ancient poetry and prose, tea had eight alternative names, such as Tu, Ming, Buyehou, Xiaoduchen, Difanzi, Qingfengshi, Yuganshi and Qingyou. Among them, Buyehou obviously say that tea has the effect of refreshment and breaking sleep. Xiaoduchen explains that tea can eliminate alcohol and botulinum toxin. Difanzi can get rid of vexation in the heart.

经典 152 茶马古道

茶马古道是指存在于中国西南地区，以马帮为主要交通工具的民间国际商贸通道，是中国西南民族经济文化交流的走廊。茶马古道是一个非常特殊的地域称谓，是一条世界上自然风光最壮观，文化最为神秘的旅游绝品线路，蕴藏着开发不尽的文化遗产。

The Ancient Tea-Horse Road refers to the non-governmental international trade channel existing in Southwest China, with the horse gang as its main means of transportation, and it is the corridor of the economic and cultural exchange of ethnic groups in Southwest China. The Ancient Tea-Horse Road is a very special regional name. It is the most spectacular natural scenery and the most mysterious cultural tourist route in the world. It contains endless cultural heritage.

茶马古道（作者：王殿维 临）

茶香图（作者：天语）

经典 153

禅茶一味

茶饮最早是在佛教寺院中广为流行，其后才传播到贵族和平民。寺院也是茶叶最早的种植基地，我国现有的茶产区几乎所有的茶叶种植都是从寺院开始的。佛教寺院在历史上还为我国培养了大批的茶学专家和人才。佛教对饮茶最为重大的影响，应该算是为饮茶注入了哲理的灵魂，使饮茶升华为一种心灵的修养和哲理的感悟，从而促成了独具东方特色的茶道的形成和发展。

Tea was first popular in Buddhist temples, and then spread to nobles and civilians. Temple is also the earliest tea planting base. Almost all tea planting in China's existing tea producing areas started from temples. Buddhist temples have trained a large number of tea experts and talents for China in history. The most important influence of Buddhism on tea drinking is that it infuses the soul of philosophy into tea drinking, and sublimates tea drinking into a kind of spiritual cultivation and philosophical perception. Thus the formation and development of tea ceremony with unique oriental characteristics was promoted.

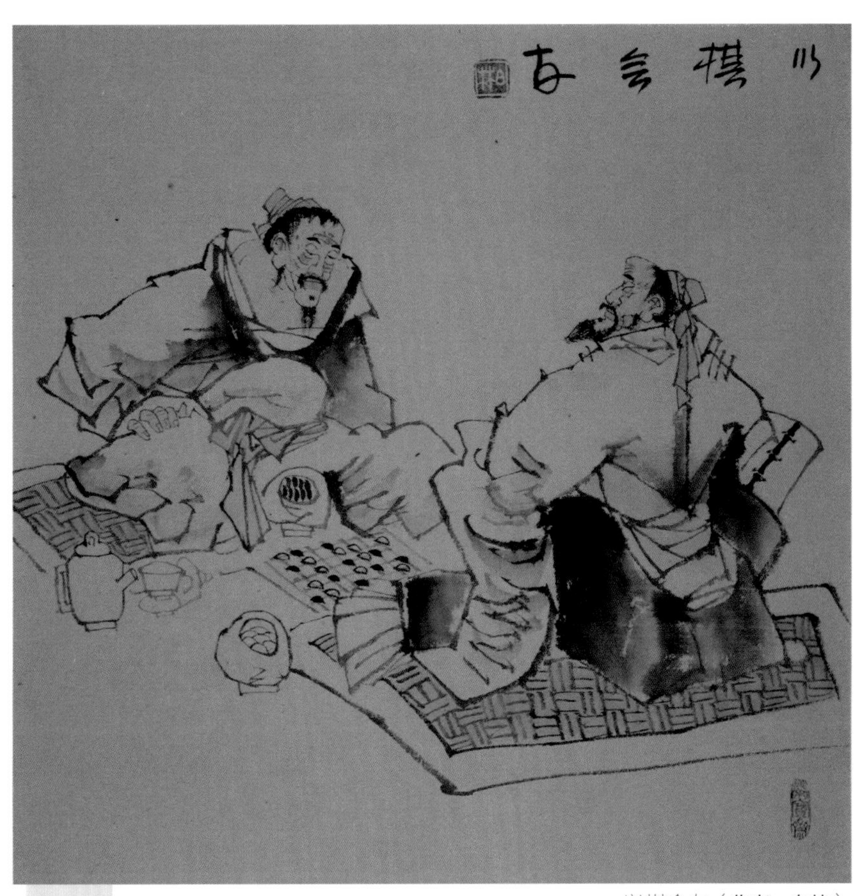

以棋会友（作者：白林）

经典 154

茶叶战争

在周重林与太俊林合著的《茶叶战争》中，作者认为，鸦片战争便是因茶而起。中国茶叶输入英国后，导致该国产生了巨大的贸易逆差。为扭转这种逆差，英国向中国输出了鸦片。茶是因，鸦片是果，鸦片的输入又导致中国白银大量外流。为了银子，中国有了禁烟运动。茶、银、鸦片的循环，最终引发鸦片战争。

In the Tea War co-authored by Zhou Zhonglin and Tai Junlin, the authors thought that the Opium War was caused by tea. After Chinese tea was imported into Britain, the trade deficit of this country was huge. In order to reverse this deficit, Britain exported opium to China. Tea is the cause, opium is the result. The import of opium led to a large outflow of silver in China. For money, China had an anti-smoking movement. The cycle of tea, silver and opium finally led to the Opium War.

经典 155 废座撤茶

“黄袍加身”的宋太祖赵匡胤召宰相范质等议政。召见之初设座赐茶，与之“坐而论道”。范质行礼毕刚要坐下，赵匡胤就说：“朕最近有些眼花，看不清东西，烦请爱卿将奏折拿到朕面前。”范质上前递折子的空当，早已受命的内侍便把宰相的座位与茶全部撤走。范质回身欲坐，发现座位与茶不翼而飞，只好站着搭话。从此，大臣们上殿议政再也不能与皇帝平起平坐 。这就是史上“废座撤茶”的典故。

清音图（作者：李子涵）

According to history, Zhao Kuangyin, the emperor Taizu of Song Dynasty, called the prime minister Fan Zhi and others to discuss politics. At first, the emperor set seats for ministers, gave them tea, and "sat and talked" with them. When Fan Zhi finished the ceremony and was about to sit down, Zhao Kuangyin said, "I've had a bit of daze lately. I can't see clearly. Please take the memorial to me." When Fan Zhi came forward to deliver the memorial, the long appointed internal servant removed the prime minister's seat and tea. Fan Zhi turned back to sit and found that the seat and tea were missing, so he had to stand and talk to the emporer. Since then, ministers can no longer sit on a par with the emporer when they go to the palace to discuss politics. This is the historical allusion of "Abolishing seat and removing tea".

经典 156

帝王茶师

在中国的历代帝王中，称得上茶学大师的非宋徽宗赵佶莫属。其所著《茶论》语言精练，见解独到。此外，他也称得上是一位真正的艺术家，能书、善画、通音律、懂鉴赏。然而做皇帝却糟糕至极，把朝政弄得腐朽黑暗，以致北宋灭亡，自己也客死他乡。

Among all the emperors in China, only Zhao Ji, Emperor Huizong of Song Dynasty, can be regarded as a tea master. The language of his book On Tea is refined and has unique views. In addition, he is also called a real artist, who can write calligraphy, be good at painting, understand music and know appreciation. However, being an emperor was extremely bad, which made the government rotten and dark, so that the Northern Song Dynasty perished, and he died in a strange land.

清隐图（作者：李子涵）

屈指叩桌

《中国民俗之谜》记载，当年乾隆皇帝下江南，带了几个太监路经松江微服来到“醉白池”游玩，在附近一家茶馆坐下来歇脚。茶房端上几只碗来，然后站在数步之外，拿起大铜壶朝碗里倒茶，茶水不偏不倚均匀地冲进碗里。乾隆看得惊奇，禁不住上前要过铜壶，学样向着其余几只碗里倒去。太监们见皇帝给自己倒茶，吓得魂都没了，想跪下叩头，叁呼万岁，又恐暴露了皇帝身份，遭杀身之祸。情急之下，纷纷“屈指叩桌”以代叩头谢礼。以后，这种谢礼的动作便在民间传开了。

赏秋图（作者：李子涵）

According to the Mystery of Chinese Folk Custom, when Emperor Qianlong went to the south of Yangtze River, he took several eunuchs along the Songjiang River, dressed in plain clothes, to "Zuibaichi" for sightseeing. They sat down in a nearby teahouse to rest. The waiter of the teahouse brought several bowls, then stood a few steps away, picked up the big copper pot and poured tea into the bowl. The tea rushed into the bowl evenly. Qianlong was surprised. He couldn't help but go up and take the copper pot and pour it towards the other bowls in imitation. When the eunuchs saw that the emperor poured tea for them, they were scared to death. They wanted to kneel down and kowtow and shout "long live", but they were afraid of exposing the emperor's identity and be killed. In a hurry, they all bowed their fingers and knocked on the table in place of kowtowing. Later, this kind of thank-you was spread among the people.

经典 158

琛瓯洗尘

知音图（作者：李子涵）

传说古代有一人名叫若琛瓯，景德镇人，以制茶具闻名。他制的茶具美观耐用，可是一个恶毒的巫师知道后，念了一道毒咒，毁坏了茶具。要解开这道咒语，需有一名年轻人投身烧茶具的炉火。为了重现精美的茶具，琛瓯勇敢地投入熊熊烈火中。咒语解开了，美丽的茶具恢复原样。人们为了纪念他，便将第一次茶水称为“琛瓯洗尘”。后人用得更多的是“洗尘”二字，以表达对客人的尊敬。

It is said that in ancient times there was a man named Ruo Chenou from Jingdezhen who was famous for making tea sets. The tea sets he made were beautiful and durable. But when a vicious wizard knew it, he read a curse and destroyed the tea set. To undo the spell, a young man needs to throw himself into the fire of the tea set. In order to make the exquisite tea set reappear, Chenou bravely put himself into the fire. The spell was undone, and the tea set was restored. In order to commemorate him, people called the first tea "Chenou dust washing". Later generations used the word "dust washing" to express their respect for the guests.

经典 159

端茶送客

品茗图（作者：李子涵）

我国古代有一种“端茶送客”的习俗。来客相见，仆役献茶，主客交谈。主人认为事情谈完，便端起茶杯请客用茶。来客嘴唇一碰杯中茶水，侍役便高喊：“送客！”主人便起身送客，客人也自觉告辞。这样的惯例，避免了主人想结束谈话又不便开口、客人想告辞又不好意思贸然说出的尴尬局面。这就是古代“端茶送客”的典故，也是茶用于古代日常交往的典型案例。

In ancient China, there was a custom of “serving tea and seeing off guests”. When the guest comes, the servant offers tea, and the host and guest talk. The host thought it was over, so he took up the tea cup to treat the guests. As soon as the guest's lips touched the tea of cup, the waiter shouted, “see off!”. The host stood up to see off the guests, and the guests also left. This practice avoids the awkward situation that the host wants to end the conversation but doesn't say it easily, and the guest wants to leave but is embarrassed to speak out rashly. This is the ancient allusion of “ serving tea and seeing off guests ”, and it is also a typical case of tea used in ancient daily communication.

经典160

茶三酒四

清茶图（作者：杨平）

“茶三酒四”是说品茶时，人不宜多，以二三人为宜。而喝酒则不然，人可以多些。这是因为品茶追求的是幽雅清静，注重细细品啜，慢慢体会。而喝酒追求的是豪放热烈的气氛，提倡大口吞下，一醉方休。这也正是茶文化与酒文化的重要区别之一。

"Tea three wine four" is to say that, when tasting tea it's not appropriate for many people, and two or three people are better. Drinking is not the case. People can have more. This is because the pursuit of tea tasting is elegant and quiet, pay attention to fine sipping, slowly experience. And the pursuit of drinking is bold and enthusiastic atmosphere, advocate swallowing, drunk. This is one of the important differences between tea culture and wine culture.

偷来浮生半日闲（作者：杨平）

经典 161

姜盐茶

传说南宋抗金名将岳飞，奉朝廷之命带兵南下与杨幺领导的农民作战。由于岳家军多来自北方中原大地，进入江南后很多士兵水土不服，出现腹胀、腹泻、呕吐、乏力症状，眼看战事难以完成。平时喜读医书的岳飞将当地盛产的茶叶、芝麻、生姜、黄豆一起熬煮让属下饮用，竟然治好了军中恶疾。此茶被称为“姜盐茶”，具有健脾胃，驱风寒，去腻强身的药用效果，被广为流传。

It is said that Yue Fei, a famous general who fought against the Jin Dynasty in the Southern Song Dynasty, was ordered by the imperial court to lead his troops southward to fight with the peasants led by Yang Yao. Because Yue warriors mostly came from the North Central Plains, many soldiers did not adapt to the local soil and water after entering the south of the Yangtze River. They suffered from abdominal distention, diarrhea, vomiting and fatigue, which made it difficult to complete the battle. Yue Fei, who likes reading medical books, cooked the local tea, sesame, ginger and soybeans for his subordinates to drink, which even cured the disease in the army. This tea is called "Ginger Salt tea", which has the medicinal effect of strengthening the spleen and stomach, dispelling wind and cold, removing greasiness and strengthening the body, and is widely spread.

经典 162

奶茶

流年（作者：宋茜炆）

唐时，文成公主和亲西藏，嫁妆自然丰厚。除金银首饰、珍珠玛瑙、绫罗绸缎外，还有各种名茶。因为文成公主平生爱茶，养成了喝茶的习惯，而且喜欢以茶敬客。初到西藏，饮食不惯，文成公主想了一个办法，先喝半杯奶，再喝半杯茶，果觉胃舒服多了。以后她干脆把茶汁掺入奶中一起喝，无意中发现混合后的茶奶，味道比单一的奶或茶更好喝。如此便相传开来，这就是最初的奶茶。

In the Tang Dynasty, Princess Wencheng married to Tibet, and the dowry was naturally rich. In addition to gold and silver jewelry, pearl and agate, silk and satin, there were all kinds of famous tea. Because Princess Wencheng loves tea all her life, she has formed the habit of drinking tea and likes to treat her guests with tea. When she first arrived in Tibet, she was not used to eating. Princess Wencheng thought of a way to drink half a cup of milk and then half a cup of tea. She felt much more comfortable. Later, she simply mixed the tea juice into the milk to drink together, and accidentally found that the mixed milk tea tastes better than a single milk or tea. In this way, it had been handed down. This is the original milk tea.

四

茶语清心

Tea, the auspicious grass with fragrance, is sweet and pleasant. Clear the mind and meditate, remove stagnation and miscellaneous dust. Not to be kitsch, not hidden in grandeur. Throughout the Chinese food culture, only this product can float in the mountains and forests, immerse in the Zen house, be active in downtown areas, and enjoy the feast in magnificent hall. It is worthy of being a national drink and a world drink.

茶，瑞草魁香，入口甘怡。清神凝思，去滞尘杂。不流于媚俗，不隐于堂奥。纵观国人饮食文化，唯此一品能飘然于山林之中，沉浸于禅房观宇，活跃于井水闹市，宴乐于朝会华堂。当之无愧为国之饮品、世界饮品。

月色未央

一杯温润漫茶香，一曲琴韵淡墨痕。袅袅茶香，一曲轻歌烂漫，一曲月色未央，时光若水，暗香浮动，细品人生味道，苦涩，甘甜，入醉。

Tea fragrance curling up, a light romantic song, a song of endless night, time is like water, and secret fragrance floats. I taste life carefully, bitter, sweet and drunk.

对弈图（作者：何奕勇）

拈花微笑，喝茶悟道。

品茗图（作者：刘士永）

经典 164

喝茶悟道

一花一世界，一叶一菩提。伴着暗香浮动，茶人们品至清至洁的茶，悟至灵至静的心。拈花微笑，喝茶悟道。

One flower one world, one leaf one Bodhi. With the dark fragrance floating, the tea people taste the clear and pure tea, and realize the spiritual and quiet heart. Picking flowers and smiling, drink tea to realize the truth.

经典165

人生如茶

茶，不过两种姿态，浮和沉。喝茶的人也不过两种姿势，拿起，放下。人生如茶，沉时坦然，浮时淡然。要学会拿得起，也要学会放得下。

Tea is just two kinds of posture, floating and sinking. There are only two positions for tea drinker, picking up and putig down. Life is like tea, calm when sinking, indifferent when floating. To learn to hold it up, but also learn to put it down.

达摩得悟图（作者：何奕勇）

品茗图（作者：尚秋香）

经典 166

茗似佳人

绿茶自然，若即若离，做情人最合适；红茶虽浓郁，任劳任怨，做主妇最好；花茶爽心，知心悦人，不妨留做知己。

Green tea being natural and neither friendly nor aloof, is most suitable to be a lover; black tea being strong, hard-working, housewife is the best; scented tea being refreshing, heart-to-heart and pleasant, you may as well stay as a confidant.

经典167 男人如茶

大本事（作者：玉砚 临）

女人似水，男人如茶，婚姻就是一个杯子。

Women are like water. Men are like tea. Marriage is a tea cup.

茶登大雅之堂而不骄淫，入茅棚草舍而无卑贱。

Tea ascends the hall of elegance without arrogance and lust, and enters the thatched cottage without baseness.

陋室铭（作者：杨平）

春水煎茶（作者：白林）

经典 169

有舍才有得

师父问：“如果你要烧壶开水，烧到一半时发现柴不够了，怎么办？”有的弟子说赶快去找柴，有的说去借，有的说去买。师父摇摇头，说：“为什么不把壶里的水倒掉一些呢？世事总不能万般如意，有舍才有得。”

Master asked, “if you want to boil a pot of boiling water and find that there is not enough firewood when it is half done, what can you do?” some disciples said to go to find firewood quickly, some said to borrow it, some said to buy it. Master shook his head and said, “Why don’t you pour some water out of the pot? Things in the world can’t be as good as they should be. Only when you give up can you get them. ”

经典 170 幸福很简单

劳动最光荣（**作者：钱松山**）

小时候觉得幸福很简单，一包糖、一个玩具就可以乐在其中。长大后发现简单很幸福，一屋一舍、一茶一饭都是其乐融融。

When I was a kid, I thought happiness was very simple. A bag of sugar and a toy can enjoy it. When I grow up, I find simplicity very happy, one room one house, one tea and one meal are all happy.

品茗毛尖（作者：尚秋香）

经典 171 淡泊平和

盈盈一盏茶，载沉载浮，却又淡泊平和，充满了生命的张力。你若问我什么是美，我会说每一盏茶都是答案。

Full of a cup of tea, carrying ups and downs, but also indifferent and peaceful, full of the tension of life. If you ask me what is beauty, I will say that every cup of tea is the answer.

经典 172

顺其自然

伴着茶香，和着琴弦，翻着古卷，恍惚间，已将日子一页一页掀过。回首仰望窗前那盆素雅的吊兰，听着花瓣飘落的声音，一丝惆怅倏然弥漫。年龄渐长，阅历渐增，习惯性的笑容里，渐渐有了些许落寞隐含其中，缘何？费尽思量，却不得而悟。罢了，顺其自然，不去追究。

With the fragrance of tea, with the strings, turning over the ancient scroll, trance, the days have been turned over page by page. Looking back at the elegant Chlorophytum in front of the window, listening to the sound of petals falling, a trace of melancholy suddenly filled. As I grow older and more experienced, I gradually feel a little lonely in my habitual smile. What's the reason? I think hard, but I can't understand. Just let it be and don't pursue it.

万物静观皆自得（作者：唐芳）

经典 173 茶水情深

携手相伴（作者：秋田）

茶从离开茶树那一刻起，就期待着与水相逢。水唤醒茶，茶成就水；水包容茶，茶激荡水；茶因水而重生，水因茶而丰润……

From the moment tea leaves the tree, it looks forward to meeting water. Water wakes tea, tea makes water; water holds tea, tea stirs water; tea regenerates because of water, and water is rich because of tea.

美好的东西常常是相通的。茶如诗词，有的婉约，有的豪放；茶如书法，有的丰润如“颜筋”，有的劲瘦如“柳骨”，有的中规中矩如隶楷，有的张扬奔放如“颠张狂素”；茶如歌，有的抒情温婉，有的激越豪迈，有的清远悠扬，有的荡气回肠。

Good things are often interlinked. Tea is like poetry, some graceful, some grandiose. Tea is like calligraphy, some are as plump as “Yanjin”, some are as thin as “Liugu”, some are as regular as Likai, some are as bold and unrestrained as “Dianzhang Kuangsu”. Tea is like a song, some are lyrical and gentle, some are passionate and heroic, some are distant and melodious, some are spirited and moving.

清供图（作者：孙卫东）

茶的信仰（作者：李国锰）

经典 175 茶的信仰

“明日隔山岳，世事两茫茫”。只有经历生命的无常，才懂得如何珍惜，如何感恩。生活依然在继续，无所谓开端，无所谓终结。茶虽不是宗教，却是我一生的信仰。

“Tomorrow you and I will be blocked by the mountains. The human relationship and world affairs are so remote!” Only through the impermanence of life can we know how to cherish and appreciate. Life goes on, no beginning, no end. Although tea is not a religion, it is my life’s faith.

经典 176

毛尖茶語
可惜一枝如畫為誰開
歲在戊戌之春於泉城
上善堂延國

一枝如画为谁开（作者：延国）

古代言茶者，莫精于陆羽。有人曾说：”饮茶人不知陆羽，如同佛子不知释迦牟尼，道者不知老庄，美国人不知华盛顿，法国人不知拿破仑。”

In ancient times, no one who talked about tea was more proficient than Lu Yu. Someone once said: “people who drink tea doesn’t know Lu Yu, just as Buddha doesn’t know Sakyamuni, Taoist doesn’t know Laozhuang, American doesn’t know Washington, French doesn’t know Napoleon.”

经典 177

喝茶要慢　做人要笨

春风拂槛露华浓（**作者：延国**）

喝茶要慢一点，做人要笨一点。慢一点喝茶才能体味到茶的真滋味，做人笨一点才能走得更踏实，看事情才能看得更明白。

Drinking tea should slow down a bit, and to be a person should be a little stupid. Drink tea slowly, then you can taste the real taste of tea. To be a little stupid is to walk more steadily and see things more clearly.

经典 178

三香萦室

乡愁自此带茶香（作者：舒肖 临）

这秋日里的夜晚，明月在窗，虫声在耳，研墨试笔，品茗夜读。书、茶、墨，香香生色，足以醉满心池。在中华传统文化里，经常把书香、茶香、墨香并提。书香悠悠，茶香逸逸，墨香淡淡，三香萦室，都是人生的美滋味、美境界。

On this autumn night, the bright moon is hanging in front of the window, and the sound of insects is ringing beside my ears. I rub an ink stick and write. Enjoy tea and read at night. Books, tea and ink are fragrant and colorful enough to make me drunk. In traditional Chinese culture, the fragrance of books, tea and ink are often mentioned together. The book fragrance is long, the tea fragrance is elegant, and the ink fragrance is light. The three fragrance lingering around the house is all the beauty taste and realm of life.

毛尖赏茶（作者：尚秋香）

经典 179

越简单越幸福

有人说，人生，一杯好茶，一本好书，一个知己，足矣。越简单越幸福。

Some people say, life, a cup of good tea, a good book, a confidant, enough. The simpler, the happier.

经典 180

让当下幸福

富贵平安图（作者：马纪奎）

再香的茶，不要隔夜。

再美的回忆，不要经年。

时时清洗茶杯，杯有清气，入茗必香。

每天清空心事，心有余闲，幸福自来。

舍不得清洗昨夜的香茗，必然喝坏今天的肠胃。

放不下既往的人事，难免有损当下的幸福。

空杯心态，让往事安眠，让当下幸福。

No matter how fragrant the tea is, don't stay overnight. No matter how beautiful memories are, don't go through the year. Clean the tea cup from time to time. There is a clear air in the cup. It will be fragrant when making tea. Empty your mind every day. Your heart has leisure and happiness comes from you.

参考文献

[1] 周国富主编:《世界茶文化大全》，中国农业出版社2019年版。

[2] 省级地方标准:《都匀毛尖茶DB52/T 433-20018》。

[3] 黔南州茶叶产业化发展管理办公室编:《都匀毛尖茶》，中国文化出版社2013年版。

[4] 杨启刚主编:《黔茶之魂》，黄河出版社2015年版。

[5] 魏明禄编著:《黔南茶树种质资源》，云南科技出版社2018年版。

[6] 魏明禄主编:《鱼钩巷》，光明日报出版社2016年版。

[7] 2017都匀毛尖(国际)茶人会首届“都匀毛尖杯”中华茶诗词大赛作品。

[8]《黔南日报》《贵州日报》。

[9] 黔南州茶叶产业化发展管理办公室“都匀毛尖茶”官网。

[10] 360百科、茶网、百度百科等。

AFTERWORD

后 记

受国学频道“用名家画笔为祖国山河立传”宣传语启发，突发奇想：既然名家画笔能为祖国山河立传，为何不能为都匀毛尖茶扬名？于是便开始收集定制相关的名家字画，整理撰写文字材料。

在中国的历史文化长河中，绘画与茶被赋予了共同的艺术审美观和生活态度。中国文人所好的"琴棋书画诗曲茶"七件事，就精神内涵而言，本身就是相通的。

编撰此书，旨在借助国内著名书画家的妙手丹青，“以茶入画，以画释茶”，面向全球弘扬中国名茶都匀毛尖茶之茶文化和茶知识，彰显茶都之地域风采，宣传毛尖茶人之奋斗历程，以期丰富都匀毛尖茶文化内涵与形式，提升影响力和美誉度，最终实现拓展市场，提振经济，造福广大茶企和茶农之目的。

好茶离不开好的制茶人，优秀的茶文化绝对不能少了优秀的文化茶人。需要说明的是，本书介绍的都匀毛尖茶茶人仅为其中杰出代表的一部分。由于时间和联系方式的限制，编撰中没能和每一位优秀茶人深入沟通和交流，有的仅从各类书刊、互联网或其他途径获得，难以全面、详实、完美地把他们的茶人精神呈现给大家。假以时日，必将弥补此遗憾。另外，英文部分是根据文本需要为外国友人了解、学习都匀毛尖茶文化所作的介绍，不完全是相应文章一对一的翻译。书中所选茶画，除一些重要章节专门联系画家朋友特别定制外，其余画品均是根据文章内容有针对性地作了选择，希望通过这些丰富多彩的艺术作品，吸引读者眼球，激发读者兴趣，加深读者对都匀毛尖茶文化乃至华夏茶文化的了解和热爱。

本书得到著名画家王殿维、单继平、王长存、贺成才、范德昌、斯曼如、舒肖、于洲桐以及著名书法家王玉柱、李超华、李传平等的大力支持。图画原作暂藏寒舍，书画图片绝大部分由中国摄影家协会吴昌琪先生友情拍摄。书中都匀毛尖茶生产工艺介绍经韦洪平、方兴齐两位茶叶专家阅审。难能可贵的是，本书文稿及出版有幸得到贵州省黔南州政协主席魏明禄、黔南州人大常委会原副主任丁匀、都匀市政协主席杨青、黔南州司法局原副局长吴贵成、都匀市文联原主席邱祥彬、都匀市书法家协会艺术顾问苏意明、都匀市教育局原副局长张承明、都匀毛尖茶文化产业协会秘书长王晓等的指导与帮助。感谢魏明禄先生为本书作序，感谢邱祥彬先生为本书编校提出的意见和建议。同时，本书的出版也离不开家人和身边亲友们的关怀与支持。这里一并表示感谢！

编者水平有限，缺点和遗憾在所难免，恳请读者朋友予以批评指正。

沈苏文

2020.08.20